Financial Liberalization and the Internal Structure of Capital Markets in Asia and Latin America

Edited by Miguel Urrutia

 THE UNITED NATIONS UNIVERSITY

© The United Nations University, 1988

The views expressed in this publication are those of the authors and do not necessarily reflect the views of the United Nations University.

The United Nations University
Toho Seimei Building, 15-1 Shibuya 2-chome, Shibuya-ku, Tokyo 150, Japan
Tel.: (03) 499-2811   Telex: J25442   Cable: UNATUNIV TOKYO

Printed in Hong Kong

DSDB-17/UNUP-638
ISBN 92-808-0638-6
United Nations Sales No. E.88.III.A.7
02500 P

CONTENTS

Introduction
Miguel Urrutia 1

On the Methodology of Comparative Studies
Kazushi Ohkawa 7

FINANCIAL LIBERALIZATION IN LATIN AMERICA

Latin America 1981–1984: Crisis, Adjustment, and Recovery
Andrés Bianchi 17

The Rise and Fall of Capital Markets in the Southern Cone
Joseph Ramos 43

The Failures of the Capital Market: A Latin American View
Eduardo Sarmiento 82

Comments on the Mexican Financial System
David Ibarra 110

A Note on Outward-looking Policy and Growth Performance:
The Republic of Korea and Selected Latin American Countries
Hirohisa Kohama 122

THE STRUCTURE OF CAPITAL MARKETS IN ASIA

Economic Development and Financial Liberalization in the Republic of Korea: Policy Reforms and Future Prospects
Kim Joong-Woong 137

Financial Opening in Malaysia
Andrew Sheng   172

Economic Growth and Regulation of Financial Markets: The Japanese Experience during the Post-war High-growth Period
Juro Teranishi   180

Financial Liberalization and the Internal Structure of Capital Markets: The Philippine Case
Mario B. Lamberte   201

The Financial System and Development: The Formation of Financial Institutions for Labour-intensive Sectors
Akio Hosono   268

Contributors   288

INTRODUCTION: EXPERIENCE WITH REGULATED FINANCIAL SYSTEMS IN ASIA AND LATIN AMERICA

Miguel Urrutia

During the time that I lived in Tokyo, while I was Vice-Rector of the United Nations University, I had occasion to follow closely the public debates concerning financial reform in Japan, and also to study and discuss with local economists the innovations being introduced in the financial markets of other Asian countries. I was surprised to find little reference to the financial liberalization disasters in Latin America. It occurred to me that some of the economists studying financial liberalization in Asia might profit from looking at what had happened in Latin America when the recommendations of the new financial orthodoxy of North American academic economists were applied. The results of financial liberalization in the Southern Cone countries suggested that the new theories made assumptions about the functioning of financial markets in developing countries which were unwarranted.

On the other hand, it appeared that the very gradual liberalization of foreign exchange transactions and financial markets in Japan was a process worth studying for Latin American economists. Furthermore, it seemed that the North American economic literature on financial liberalization in the Republic of Korea, an experience which was used by foreign advisers working in Latin America as an example of how the elimination of financial repression could promote economic growth, simply did not reflect accurately Korean economic history. It was useful, therefore, for Latin American economists to find out, without intermediaries, what had been happening in some of Asia's most successful economies.

When I explored with the authorities of the United Nations Economic Commission for Latin America and the Caribbean (ECLAC) whether they would be interested in discussing the issue of financial liberalization with Asian scholars, I found an enthusiastic response, for two reasons:

1. It appeared that the economists of ECLAC had never before had the occasion to participate in a conference or workshop that analysed Asian economic experiences, and;
2. ECLAC was in the process of evaluating the benefits and costs of the recent financial liberalization episodes in the region.

Given this positive response, and with the support of Rector Soedjatmoko of the United Nations University, I presented the idea of holding a Latin American–Asian workshop on financial liberalization to Professor Kazushi Ohkawa, of the International Development Center of Japan (IDCJ). He gave support to the initiative, and we contacted a group of Asian scholars who were known to be working on issues of financial development.

The present volume presents some of the papers discussed at the workshop. The workshop participants were very stimulated by the intercontinental comparisons, and unanimously agreed that it would be useful if these papers were published, so that a wider audience might be made aware of economic experiences only very imperfectly known in each others' regions.

In what follows, I will attempt to reflect some of the issues that came up in the discussions.

The Methodology of Comparative Studies

The paper by Professor Kazushi Ohkawa addresses the methodological problems of economic comparative studies. On the basis of his wide experience in the international comparison of the growth experience of different nations, he sets out the usefulness of carrying out comparative studies within the perspective of the different development phases which most nations that developed after 1870 have passed through. He discusses the apparent differences between Asian and Latin American development by suggesting that one particular phase of development crucial in East Asia, that of primary export substitution, seems to have been skipped or postponed in Latin America, and that this may explain many of the differences in development between the two regions.

He also explores some of the causes for differential industrial performance in Asia and Latin America. His empirical finding that the residual, which might be a proxy for technological-organizational progress, is much lower in Latin America than in Asia opens up a whole new area for further research.

The description of the different development phases in East Asia also suggests that financial liberalization was not carried out successfully before the phase of secondary import and export substitution. Since few countries of Latin America have fully entered into that phase, the failures connected with the financial liberalizations of the 1970s are not inconsistent with the Asian experience.

Professor Ohkawa's paper points out the importance of having a common framework in comparative studies in order to maximize their usefulness for the building of economic theory and for policy analysis. For this seminar a looser framework than that proposed by Professor Ohkawa was used, and the contributors were only asked to present the experience of financial liberalization in their country. From reading these papers it becomes clear that further comparative analysis on the role of the financial sector in different phases of development would be very interesting, and that probably different phases of development call for different degrees and types of financial regulation.

The Latin American Experience

The paper by Andrés Bianchi describes the crisis of the 1980s in Latin America, and serves as background for the discussion of the Latin America experiments with financial liberalization. Such background is crucial for non-Latin Americans to understand the multi-faceted nature of this crisis. Although financial liberalization may have helped to create the crisis in some countries, there are other important causes of economic stagnation and even of the financial crises experienced in 1981–1984.

The paper by Hirohisa Kohama also serves as background on Latin America for Asian readers. It compares the larger Latin American economies with the Republic of Korea, utilizing the "economic phases" categories discussed by Professor Ohkawa.

The paper by Joseph Ramos is a thorough, illustrative survey of the financial liberalization experiments in Argentina, Brazil, and Uruguay in the 1970s. His final assessment is that the experiments did not live up to the expectations of their authors. He states that "the profound changes which financial liberalization and opening up brought about did not translate themselves, despite intentions, into systematically higher savings or into clearly improved resource allocation. Indeed, the three experiences came to a close with their financial systems in a shambles."

His analysis suggests that part of the failure of the financial liberalization attempts was due to the economic setting in which the reforms were carried out. He concludes that "it was thus a grave policy error to have liberalized financial markets so rapidly, and to such an extent, precisely at a time when, because of the stabilization policy, important disequilibria still remained to be resolved in other critical sectors of the economy."

In addition to the problems of trying to liberalize financial markets at a time when major disequilibria persist in other sectors of the economy, Ramos also thinks that the architects of the reforms did not take into account the inevitable increases in consumption that liberalization would inevitably bring about. He explains this point in section 7 of his conclusions. "Also overlooked by most policy-makers was the fact that 'financial repression' not only kept interest rates artificially low but, by rationing credit, necessarily repressed the demand for certain types of credit (generally that for consumption). It was thus a serious oversimplification for neo-conservative theorists to focus exclusively on the favourable effects financial liberalization might have on effective savings and investment (via higher interest rates) and to neglect the unfavourable effect it could have on these by releasing the pent-up demand for consumption."

Eduardo Sarmiento makes the same point in his paper, but goes further in the critique of those who argued for liberalization as a way of increasing savings and improving resource allocation. The very title of his paper suggests that there are substantial imperfections in the capital markets of developing countries which would argue against liberalization. The experience in Latin America in the 1970s gives strong empirical backing to this thesis.

In summary, the description of the financial liberalization experiments in Latin America in the 1970s suggests that the results that the policy-makers expected in terms of savings, interest rates, and efficient allocation of credit did not become a reality. Clearly some of their assumptions concerning the functioning of the financial system had not been realistic.

The paper by David Ibarra is somewhat different, and its value is due not so much to the analytical contribution it makes to an understanding of the Mexican case as to the fact that it is a description of events from the point of view of one of the major actors in a crucial period of Mexican economic history. David Ibarra was the Mexican Minister of Finance (1977–1982) during the period in which oil began to dominate the Mexican economy and in which the Mexican debt started to grow rapidly. The paper gives some hints about the political determinants of policies that in hindsight seem to have led Mexico to the debt and inflation crisis of 1982–1986. The discussion of this paper at the seminar was particularly interesting, in so far as it illustrated the role of the different social sectors in decision-making in Mexico. One impression one got was that the extensive negotiations with various social groups leading to the important tax reform which introduced the value added tax took up a good part of the time of the policy-makers, while major unexpected shifts were taking place in the Mexican economy owing to oil-price changes and public sector investment decisions that seemed outside the control of the economic authorities.

The paper by David Ibarra suggested to the participants the importance of studying the process of political decision-making in different societies in order to understand the process of financial development.

The Asian Experience

The Asian papers were a surprise for the Latin American participants. Many of the foreign advisers on capital market development who had come to Latin America had argued that some of the economic success of East Asian economies was due to the adoption of policies of financial liberalization.

The paper on the Republic of Korea, and discussions with Kim Joong-Woong, suggested that in fact the financial markets in that country were highly regulated. Although a liberal economist who favours some further liberalization of financial markets in the Republic of Korea's present stage of development, Kim maintains in his paper that, "in the earlier stages of development, the use of direct and selective instruments of financial policy ensured an efficient allocation of financial resources." He then describes financial practices not unlike those which receive so much criticism from foreign experts who advise Latin American governments: "The Korean financial sector has been effective in channelling financial resources to the preferred or strategic sectors designated in the economic development plans. This was achieved mainly through preferential loans extended at low interest rates." He then explains that "the major commercial banks were all largely public enterprises... with the result that there was direct government control of virtually every aspect of banking operations. Second, as a means of allocating investment funds according to national priorities, the government resorted to strict control of interest rates and provided extensive preferential credit at subsidized rates." Clearly, the high Korean growth rates are not explained by the existence of liberalized financial markets.

The country where liberalization has gone further has been Malaysia, but even there freedom in financial markets does not determine higher savings rates. Andrew Sheng points out that a significant portion of the high savings rates (30 per cent of GNP) was captive (forced savings) in the form of provident fund contributions. Furthermore, interest rates were freed only in 1978, and lending guidelines to priority sectors were maintained.

In Japan, state intervention in financial markets has also been extensive. Akio Hosono describes the institutions created by the state to meet the credit needs of different sectors of the economy that would not normally have access to funds in the free capital market.

Professor Juro Teranishi introduces his paper *Economic Growth and Regulation of Financial Markets* with a rather provocative statement: "What was particularly interesting for researchers in money and banking was that this rapid growth [9.2 per cent GNP growth rate between 1953 and 1972] had taken place within a highly regulated financial framework."

An interesting point underlined in the Teranishi paper is that the strict regulation of international capital transactions seems to have provided the necessary conditions for sustaining the regulated financial system. The Latin American experience described by Ramos, Ibarra, and Sarmiento would suggest the same hypothesis. In Mexico, where international capital transactions cannot be easily regulated, not too much regulation in the financial system is possible. In Colombia, foreign exchange regulations that control with some degree of effectiveness international transactions have sustained a regulated financial market during different phases of development.

Finally, the paper by Mario B. Lamberte on the Philippine financial market points out again the surprising similarity in policy framework existing in Latin America and the Philippines.

Conclusions

In summary, the participants came away from the cross-continental exchange with the feeling that much of the recent theory pointing out the advantages of financial liberalization may overstate such advantages and ignore some of the realities which may make rapid liberalization a costly endeavour for a developing country. In particular, it would seem that the combination of trade and financial market liberalization may be quite explosive, and lead to a rapid increase in consumption of durable consumer goods, and this might have very dangerous effects on savings and the balance of trade.

On the other hand, even the most outspoken critics of the advocates of financial liberalization were careful to point out that maintaining interest rates at unrealistically low levels served no useful development purpose. Intervention in financial markets must be limited and realistic to be effective.

In general, the analysis of the financial liberalization case studies made all of the participants very aware of the need to proceed very gradually down this road. When the conference was organized there was little expectation that Asian and Latin American scholars would find themselves so easily in agreement in this policy area.

ON THE METHODOLOGY OF COMPARATIVE STUDIES

Kazushi Ohkawa

Introduction: The CA Project*

The major result of the CA project, "Japan's Historical Development Experience and the Contemporary Developing Countries: Issues for Comparative Analysis" has recently been published.[1] I should like to quote the following passage from its Introduction:

It is our conviction that, in the pursuit of a better understanding of what development is all about, the *comparative historical laboratory* has been seriously underutilized, especially relative to multi-country cross-sectional efforts, on the one hand, and strictly contemporary or forward-looking country-intensive efforts, on the other. The entire CA project has been based on the premise that analysis of past development experience, particularly of a generally "successful" case such as Japan, has something to teach us not so much in terms of the precise transferability of a total societal experience, as something about the transferability and, at times, non-transferability of a number of special sectoral or function dimensions considered important to an explanation of the Japanese experience [my italics].

---

*CA is an acronym for a comparative analysis project entitled "Japan's Historical Development Experience and the Contemporary Developing Countries; Issues for Comparative Analysis," which was carried out by the International Development Center of Japan (IDCJ) from 1975 to 1981, in collaboration with the Economic Growth Center of Yale University. The purpose of the project was to gain a clearer understanding of Japan's historical development pattern in relation to major critical issues in contemporary LDCs' growth performance.

In fact, the project has not been in a position to compare the Japanese historical experience in every dimension. In the selection of research topics and the ensuing results presented in the published volume, the areas focused on include an examination of the relevance of initial conditions and an overview of the historical experience (part I), the particular importance of technology choice and change in agriculture (part II) and in industry (part III), the flow of resources between the agricultural and non-agricultural sectors, the role of institutions and policies in finance (part IV), and the relation between trade and development (part V). The project was also not in a position to compare the Japanese experience with representatives of every other typology of contemporary developing countries, where typologies are defined by different clusters of characteristics such as initial conditions, resource endowment, country size, etc. The countries chosen for comparison are Thailand (initial condition), Republic of Korea, and Taiwan (historical perspective) in part I; Republic of Korea, Taiwan, and the Philippines, Indonesia and Colombia in part II; India (cotton) and Brazil (steel) in part III; Taiwan (terms of trade), India (resource flow), and the Philippines (finance) in part IV; Thailand (textile industry), East and South-East Asia (automobile parts production) and Republic of Korea, Thailand (trading companies) in part V. To our regret, the CA project did not include the study area which is directly relevant to the theme of this symposium, "comparative analysis of the experience of financial liberalization in Asia and Latin America."

Thus the project fell short in its capacity (financial and human) to undertake comparative analysis in a comprehensive way both across major typologies (particularly the Latin American and African cases, which are more "distant" from the Japanese model), and across countries within "neighbouring" typologies, for example, the South-East Asian and South Asian cases. In retrospect, ours should be called a selective, partial, comparative studies methodology.

Typologies and Phases

In the light of the limited experience of the CA project research, what I would like to discuss is the relation between typologies and development phases, as I believe this is a basic problem for comparative historical analysis.

In presenting an overview of the historical experience of Japan's development, we demarcated several development phases largely in conformity with the shorter post-war experience of Taiwan and the Republic of Korea: P1, 1870–1900, traditional export expansion, coupled with primary import substitution; P2, 1900–1920, primary export substitution; P3, after 1920, secondary import and export substitution. P1 is 1950–1962 (Taiwan) and 1953–1964 (Korea); P2 is 1962–1970 (Taiwan) and 1964–1972 (Korea); P3 is after 1970 (Taiwan) and 1972 (Korea).[2]

I should like to quote the paragraph relevant to our methodology for deriving such phases.

Analysis of comparative growth suggests sets of countries can be grouped around basic family affinities, a certain uniqueness not necessarily shared by other types of developing countries. However, even within a family of less developed countries (LDCs), there may exist important, and instructive, differences. Recognition of such differences as well as similarities in behaviour among developing economies permits the generation of a more flexible evolutionary view of development, based on the notion that each phase in the transition (in Kuznets' sense) is characterized by a distinct structural form (morphology) and distinct mode of operation (physiology). Such phases may be identified through a combination of inductive evidence and deductive reasoning.

Can the phasing of historical experience of East Asia be applicable to Latin America? The answer seems to be "no" for a number of authors. The major difference might be the "skipping" of P2, the phase of primary export substitution (substitution of exports of agricultural products by those of manufactured goods), and a more prolonged P1 in most of the Latin American countries.[3] The phenomenon has been noted also to a certain extent in some of South-East Asia, for example Thailand and the Philippines. Although our research in the CA project did not cover this area, as has been pointed out earlier, we believe such differences can no doubt be identified.

The problem is "how to interpret it." It seems to me that the phase performance of East Asia has often been taken as the standard type, against which other types of phase performance, such as "skipping of phase 2," are to be evaluated. The assertion is made, when analysing the development strategy, that prolonged import substitution policies, including various kinds of protection measures, might be responsible for bringing about a behaviour which is "distorted" if compared with the "assumed" standard performance of phase sequence identified for the historical experience of Japan and other East Asian cases. I do recognize the difference of policy option and share a view that this has been influential in forming the different phase behaviour. However, it is my view that as a methodology of comparative analysis, we had better pay much more attention to the typological difference itself before going on to the policy debates, and this is the route we have taken in carrying out the CA project.

The typological difference, as seen broadly between East Asia and Latin America, is primarily rich $v.$ poor natural resource endowment, among other things, although the implications are different when the country is rich in agricultural land or in mineral resources. Such a typological difference is also evident to a certain extent within Asia, say between East Asian and South-East Asian countries. Even within East Asia, Taiwan is relatively rich in agricultural resources. We should also discuss the difference between Latin American countries in this regard. Nevertheless, it can be assumed that we will be able to classify all the countries into groups of similar typology based on their affinity. In the present context, we are concerned with the East Asian, South-East Asian, and Latin American groups in particular. It would be quite natural to think about the possible variances in industrialization performance due to the typological difference, which economically is in terms of comparative advantage of primary sectors. The reason for describing such a seemingly obvious fact stems from our view that there may be an alternative approach which avoids the concept of standard type of phase performance. This alternative approach would interpret all the three variants as "neutral" without setting any standard type as an evaluation criterion.

My factual knowledge is limited, but it can be conjectured that each group of affinity can be characterized respectively by: (a) the pronounced and distinct performance of primary export substitution activities (East Asia); (b) the long-lasting import substitution activities (Latin America), without making a distinction between so-called non-durable consumer goods and durable goods for producers and consumers; and (c) a behaviour of intermediate type (South-East Asia). Much has been said about the East Asian and Latin American models, but the intermediate type would need a brief illustration. For example, in the Indonesian development plan and economic policies, it is hard to draw a sharp distinction between the primary and the secondary import substitution approaches at present, although certainly the primary one did start earlier and, as a matter of fact, primary export substitution is of great importance. Such a mixed type of behaviour is also witnessed, though to a lesser extent, in Thailand. The point here is that primary export substitution and secondary import substitution cannot be considered as a distinct alternative option either at the policy level or in the actual process of industrialization.[4]

Productivity and Employment

If we observe the process of industrialization of LDCs using a broader common indicator, say, the historical changes in the manufacturing output's share of GDP, irrespective of its composition in non-durables and durables, a fairly regular performance would be identified through the three groups of different typologies presented above. For example, in Japan's case, its percentage share is 10–17 in P1, 17–23 in P2 and over 23 in P3. As is widely known, the fast pace of increase in this share is characteristic of the development process and it becomes distinctly slower in the industrialized state. Therefore, comparative studies of two countries belonging to different typology groups cannot be carried out effectively without taking this aspect into account. The Republic of Korea and Brazil can be compared as representative of East Asia and Latin America respectively, but Indonesia and Mexico should not be compared. In the former case, the share of manufacturing output is more or less the same, but in the latter case the share of Indonesia is definitely smaller than that of Mexico.

The discussion leads us to a broader aspect of comparative analysis, which concerns the changes in the industrial structure in terms of domestic production. The phase demarcation in terms of import and export developed for Japan and other East Asian countries, as has been pointed out in the introduction to our paper in the CA volume, referring to C. Clark and S. Kuznets, is nothing but a deeper observation of changes in terms of comparative advantage in open economies, reflecting changes in final demand and productive capacity during the development process. With respect to the alternative approach suggested previously, I want to draw your particular attention to this aspect, before getting into the policy aspect of the problem.

In evaluating this broad process of industrialization, we believe two criteria are basic, and they are the pace of productivity and employment increase, in particular in the industrial sector. (From a social point of view, one should consider whether the latter improves income distribution or not, but I will not discuss this aspect here.) Unfortunately, our empirical knowledge about the historical performance of productivity and employment and their relationship has not been arranged so as to be able to evaluate the varied behaviour of the three groups demarcated by typological characteristics in this respect. Of course we are given a number of partial data. For example, one author points out that the incremental/capital output ratio, an indicator of capital productivity, appears to be higher in Latin American countries than in Asian countries. Another says that the employment-absorbing power of the industrial sector in the former is much smaller than the latter. These may or may not be valid statements, depending on the time.[5] At any rate, the challenge for us is, first, a problem – "how do we make our factual knowledge more systematic?" and, second, "how can we link these features, if any, to the typological characteristics or phase differences?"

I am not qualified enough to answer the challenge even in part. Nevertheless, I will try to suggest, by illustration, some of the examples of analysis towards which we have to proceed.

The incremental capital/output ratio, in its reciprocal ($\Delta Y/I$, Y: output, I: investment), can be broken down into two terms: contribution of the conventional input, capital, K, and labour, L, in their incremental, ΔK and ΔL, and the residual, R, contribution of the non-conventional factors, also in its incremental term, ΔR. If we assume the contribution of conventional factors can be evaluated by their opportunity costs (wage rates, w, and rates of capital return, r) in the base year for the hypothetical case of no technological-organizational changes in the year to be compared, we can have a simple formula as follows.

$$\Delta Y/I = \Delta R/I + w\Delta L/I + r, \qquad (1)$$

where for simplicity's sake I is used in place of ΔK, ignoring replacement investment. This is an incremental version of the usual growth accounting

Table 1. Decomposition of incremental capital-output ratio: industrial sector, Japan

Phases	I/ΔY	ΔY/I	wΔL/I	ΔR/I	I/Y	ΔR/Y	ΔY/Y
P1	3.32	30.1	4.8	13.3	16.5	2.19	5.29
P2	3.16	31.6	5.9	12.7	21.3	2.92	7.32
P3a	3.18	31.5	4.5	13.0	18.4	2.76	5.08
P3b	2.95	33.9	8.5	12.3	35.4	4.70	14.60

a. Pre-war
b. Post-war. r is flatly assumed to be 12 per cent.
Sources: K. Ohkawa and N. Takamatsu, "Capital Formation, Productivity and Employment: Japan's Historical Experience and Its Possible Relevance to LDCs," IDCJ Working Paper Series, no. 26 (Tokyo, 1983); K. Ohkawa, "Capital Output Ratios and the 'Residual': Issues of Development Planning,"*IDCJ Working Paper* Series, no. 28 (Tokyo, 1984).

formula, and is convenient for observing the behaviour of productivity and labour employment in their relationship in the evolutionary path of development.

Table 1 summarizes the result of its tentative application to the industrial sector of Japan.

I do not intend to discuss in detail the procedures and results of this hypothetical calculation. What deserves noting here is as follows:
1. The capital-output ratio tends to decrease in the long run.
2. The incremental term of Lw/K does not show a trend of decrease – instead, in the post-war P3, the phase of secondary export substitution, it increased.
3. The incremental increase of the residual per investment has been kept almost unchanged. Because of a trend of increase in the investment proportion, its value per output had increased in particular in P3b. Thus the proportion of the residual increase to the rate of output increase amounted to 30–50 per cent throughout the entire period.[6]

The residual is, as a matter of fact, a measure of "unknown" parts of output growth. Nevertheless, it does include as a major component the result of technological-organizational progress. While realizing such a sustained behaviour of productivity increase, the employment effect of investment shows a notable performance. I would draw your particular attention to the large value of P3b. It is my view that this was brought forth by a rapid growth of the machinery industry in a favourable international environment and that, speaking in general, the secondary export (and possibly import) substitution process does not necessarily have to be unfavourable in terms of solving the employment problem. Actually, in Japan's case the capital labour ratio, K/L, tends to be almost equal between the textile and machinery industries (except transportation machinery), and much lower than that of the so-called heavy chemical industries. What differs between textile and machinery is the high wage-rates of the latter, indicating a higher quality level of labour.

From a methodological point of view, it is desirable to make a comparative analysis by applying this formula to the representative countries, at least, of the

Table 2. Decomposition of incremental capital-output ratio: macro-level, selected countries (hypothetical calculation for 1960–1980)[a]

	I/ΔY	ΔY/I	wΔL/I	ΔR/I	I/Y	ΔR/Y	ΔY/Y	ΔR/Y
Brazil	3.40	29.4	8.7	8.6	22	1.9	7.05	27.0
Mexico	3.85	27.0	7.1	7.9	24	1.9	6.2	30.6
Colombia	4.20	23.0	8.1	2.9	23	0.6	5.5	10.9
Republic of Korea	2.65	37.7	7.2	18.5	24	4.4	9.05	40.6
Malaysia	2.90	34.4	8.3	14.1	21	2.9	7.15	40.3
Thailand	3.20	31.3	6.3	13.0	24	3.1	7.8	39.9
Philippines	3.90	25.6	6.3	7.3	22	1.6	5.7	28.1

a. B, labour's relative share, is flatly assumed to be 60 per cent to derive wΔL/I by use of BΔL/L, and r is assumed to be 12 per cent.
Source: Mostly World Bank, World Development Reports.

three groups of different typologies. However, available data are limited at present, and it is impossible to do so even for the industrial sector for which the application of this formula is most desirable. I hazarded a very rough measurement by using the same simple formula for some selected countries at the macro level (table 2).

Even with such a crude hypothetical/measurement, I hope it will be possible to illustrate some important characteristics of these selected countries' growth performance.

1. The magnitude of the employment term as well as the investment proportion to GDP differs little among these countries. What differs is the magnitude of the residual increase.
2. A fairly wide range of incremental capital/output ratio and the variance of output growth rate can largely be explained by the difference of the residual performance.
3. The same thing can be stated by saying that the rate of output growth due to the conventional inputs appears to be not much different: Latin American group 4.3–5.1, East Asian group (though only the Republic of Korea) 4.7, and South-East Asian group 4.1–4.7.

The rate of technological-organizational progress seems to be again at issue here, although the available data do not permit us to observe the changes in the residual growth by phase sequence. Either by investment or per output, the magnitude of incremental increase of the residual is outstanding in East Asia; the Latin American group is on the lower side of the range, while the South-East Asian group, except the Philippines, comes in between. Can we link this variance directly with their typological difference? I hesitate to say "yes." Of course, one cannot deny the validity of our historical knowledge that protective policies and hence limitation of competitive activities would discourage the realization of evolutional potentials. Policy options along this line might have retarding effects on the rate of technological–organizational innovations. However, I am much more inclined to say that the problem of technological–organizational progress, in particular for late-comers, bears on (or depends upon) a much wider range of factors, such as government policies and private enterprisers' behaviour directly related to technological changes and organizational set-ups. This is an important problem area that should be investigated further in comparative studies.[7]

Notes

1. K. Ohkawa and G. Ranis with L. Meisner, eds., *Japan and the Developing Countries* (Basil Blackwell, London, 1985).
2. J.C.H. Fei, K. Ohkawa, and G. Ranis, "Economic Development in Historical Perspective: Japan, Korea, and Taiwan," in K. Ohkawa and G. Ranis (note 1 above), p. 37. Instead of single years, the dating implies a band of years.
3. For example, in Akio Hosono, *The Economies of Latin America* (in Japanese), Tokyo University Press, 1983. He says "in Latin America, without having the phase of primary export substitution phase, industrialization made a shift directly from the phase of primary import substitution to the phase of import substitution of intermediate goods and capital goods (this is called secondary import substitution) and thus resulted in maintaining prolonged import substitution industrialization" (p. 111).
4. In Japan's case, at first glance a similar phenomenon seems to have taken place in the pre-war part of P3, between 1920 and 1940. According to our demarcation, this is the former part of secondary import substitution, and as a matter of fact this activity had been almost completed by 1940, increasingly becoming the dominant function of the economy. However, the expansion of textile exports, the core of primary export substitution, still went on actively for the whole of this period, and never stopped. The point of our phasing is that during P2 the latter activity was dominant, accompanying an early start of import substitution of machinery and other durable goods. The performance of Thailand and Indonesia appears to be different from that of Japan in that no particular phase P2 can be distinguished clearly.
5. For example, it has often been said that Taiwan and the Republic of Korea could solve the employment problem towards the end of P2 as the real wage of unskilled labour tends to increase owing to the adoption of labour-intensive industries in P2. I do share this view but consider that two additional conditions should be taken into account. First, before the beginning of the 1970s the international environment was very favourable for export expansion and the rate of export growth contributed much to the greater rate of employment increase in the industrial sector. For Japan, the bonanza of the First World War played a similar role. Second, the most important, fast productivity increase was an indispensable factor in realizing such achievements.
6. Actually, r is greater for earlier phases, so that the magnitude of $\Delta R/I$ tends to be smaller for P1 and P2.
7. Some selected aspects have been discussed in *Towards New Forms of Economic Co-operation between Latin America and Japan* (ECLA/IDCJ, Tokyo, 1980).

Financial Liberalization in Latin America

LATIN AMERICA 1981–1984: CRISIS, ADJUSTMENT, AND RECOVERY

Andrés Bianchi

Economic Crisis (1981–1983)

Dimensions

Between 1981 and 1983 Latin America experienced its severest, most widespread and longest economic crisis since the ill-fated years of the Great Depression.

Seen in historical perspective, one of the outstanding and unusual features of the crisis was the large number and varied types of the region's economies which suffered its effects. While some countries fared better than others, the decline in economic activity and the aggravation of the problems of the external sector were widespread, affecting both large economies like Mexico and Brazil – that rank among the 12 biggest in the world – and the tiny countries of Central America and the Caribbean; oil exporters like Venezuela and economies totally dependent on imports of petroleum like Uruguay and Paraguay; countries pursuing relatively dirigist and inward-oriented development strategies as well as those relying on more market-oriented and outward-looking policies.

Another unique and no less disturbing characteristic of the crisis was the generalized and simultaneous deterioration of virtually all main economic indicators.

Thus the growth rate of GNP fell sharply in 1981, causing the first decline in Latin America's per capita income since 1949. This was followed in 1982 by a reduction in the absolute level of economic activity – the first ever registered in the post-war period – and by a more pronounced decline of total output in 1983 (see table 1 and figure 1).

Table 1. Latin America: Main economic indicators

Indicators	1975	1976	1977	1978	1979	1980	1981	1982	1983	1984
Gross domestic product at market prices (billions of dollars at 1970 prices)	257	271	285	298	318	336	341	338	327	338
Population (millions of inhabitants)	302	310	318	326	334	342	350	358	367	375
Per capita gross domestic product (dollars at 1970 prices)	849	875	897	916	953	982	975	943	893	901
Per capita gross national income (dollars at 1970 prices)	848	875	898	910	951	985	962	912	860	866
					Growth rates					
Gross domestic product	3.6	5.7	5.1	4.7	6.5	5.6	1.7	−1.0	−3.1	3.3
Per capita gross domestic product	1.1	3.1	2.5	2.2	3.9	3.1	−0.7	−3.3	−5.3	0.9
Per capita gross national income	−0.5	3.2	2.6	1.3	4.5	3.5	−2.3	−5.3	−5.7	0.7
Consumer prices	57.8	62.2	40.0	39.0	54.1	56.5	56.8	84.5	130.8	175.4
Terms of trade (goods)	−13.5	4.4	6.0	−10.5	4.2	4.1	−9.2	−8.8	−6.2	0.2
Current value of exports of goods	−7.8	16.4	19.3	7.5	34.6	29.4	7.3	−7.9	−0.2	0.2
Current value of imports of goods	6.5	3.8	15.0	13.9	25.8	32.4	7.8	−19.9	−28.6	4.0
					Billions of dollars					
Exports of goods	33.6	39.1	46.7	50.2	67.5	87.3	93.8	86.4	86.2	95.0
Imports of goods	39.2	40.7	46.9	53.4	67.1	88.9	95.8	76.7	54.7	56.9
Merchandise trade balance	−5.6	−1.6	−0.2	−3.2	0.4	−1.6	−2.0	9.6	31.5	38.1
Net payments of profits and interest	5.6	6.8	8.2	10.2	13.7	18.0	27.7	37.6	34.5	37.4
Balance of current account	−14.0	−11.0	−11.8	−18.3	−19.6	−28.1	−40.6	−40.7	−9.0	−2.1
Net movement of capital	14.2	17.8	17.0	26.0	28.6	29.7	37.8	19.2	4.4	12.4
Trade balance	0.2	6.8	5.2	7.7	9.0	1.6	−2.8	−21.4	−4.5	10.0
Total gross external debt	76.2	99.7	118.8	150.9	182.0	221.0	275.4	315.3	340.9	360.2

Source: ECLAC, on the basis of official figures.

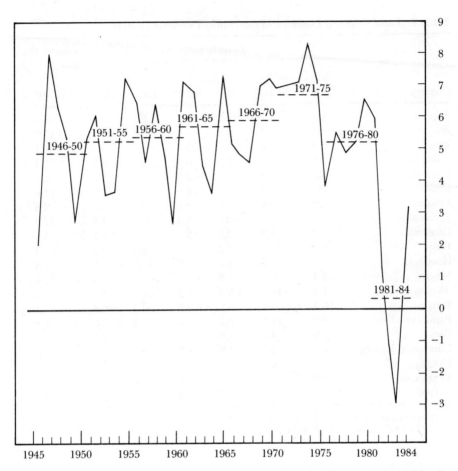

Fig. 1. Latin America: Annual growth rates of Gross Domestic Product (after ECLAC, on the basis of official information).

As a result of this downturn in economic activity and the rapid rate of population growth, real product per capita fell by almost 10 per cent between 1981 and 1983, dropping to a level similar to that reached already in 1976. This decline was, moreover, quite widespread, affecting 17 out of the 19 countries with comparable data (see tables 2 and 3).

Since at the same time the terms of trade deteriorated sharply and factor payments soared, national income – by far a better measure of economic welfare than GDP – fell even more dramatically.

As was to be expected, the slump was also especially acute in the case of investment. In fact, after growing at a strong pace in the 1970s, gross fixed capital formation stagnated almost completely in 1981 and fell by about 40 per

Table 2. Latin America: Evolution of total Gross Domestic Product

Country	Growth rates						Cumulative rate
	1975–1978	1979–1980	1981	1982	1983	1984[a]	1981–1984[a]
Argentina	4.8	3.7	−6.2	−5.1	3.1	2.5	−6.0
Bolivia	5.1	1.2	0.7	−6.6	−8.6	−3.7	−17.2
Brazil	6.5	6.8	−1.6	0.9	−3.2	4.5	0.4
Colombia	4.9	4.7	2.3	0.9	1.0	3.0	7.4
Costa Rica	5.7	2.8	−2.3	−7.3	2.3	5.0	−2.7
Chile	1.7	8.0	5.5	−14.1	−0.7	5.9	−3.8
Ecuador	7.0	5.1	3.9	1.8	−3.3	2.0	4.5
El Salvador	5.5	−5.3	−8.3	−5.6	−0.7	1.5	−12.8
Guatemala	5.5	4.2	0.7	−3.5	−2.7	0.2	−5.5
Haiti	3.7	7.5	−2.8	−2.5	1.3	2.7	−1.5
Honduras	5.8	4.7	1.2	−1.8	−0.5	2.8	1.6
Mexico	5.3	8.8	7.9	−0.5	−5.3	3.5	5.1
Nicaragua	1.2	−10.0	5.3	−1.2	4.7	−1.4	7.4
Panama	3.5	9.7	4.2	5.5	0.4	−0.8	9.5
Paraguay	9.2	11.4	8.7	−1.0	−3.0	3.0	7.4
Peru	1.5	4.0	3.9	0.4	−10.8	4.5	−2.8
Dominican Republic	4.7	5.3	4.0	1.7	3.9	0.6	10.6
Uruguay	4.1	6.0	1.9	−9.7	−4.7	−2.0	−13.9
Venezuela	5.9	−3.4	−0.3	0.7	−4.8	−1.7	−6.0
Total	4.8	6.1	1.7	−1.0	−3.1	3.3	0.8

a. Provisional estimates subject to revision.
Source: ECLAC, on the basis of official figures.

cent in 1982–1983, thereby reducing the investment coefficient from slightly over 22 per cent in 1980 to 16.5 per cent in 1983, its lowest level of the entire post-war period. Hence, in addition to depressing current living standards, the crisis had the effect of limiting the possibility of raising them rapidly in the future.

The slowdown in economic activity went also hand in hand with a sharp rise in open urban unemployment and an increase in various forms of underemployment. As can be seen in figure 2, the proportion of urban workers without jobs rose in most countries for which relatively reliable data are available and reached extremely high levels in Chile, Nicaragua, and Uruguay.

The negative social effects of the worsening in the employment situation were compounded in a number of countries by drastic falls in real wages. For example, in 1981–1982 real wages dropped by 30 per cent in Costa Rica and 20 per cent in Argentina and in 1983 they experienced a 20 per cent reduction in both

Table 3. Latin America: Evolution of per capita Gross Domestic Product[a]

Country	Dollars at 1970 prices					Growth rates				Cumulative rate
	1970	1980	1983	1984[b]	1980	1981	1982	1983	1981[b]	1981–1984[b]
Argentina	1,241	1,334	1,166	1,177	-0.9	-7.7	-6.6	1.4	0.9	-11.8
Bolivia	317	382	295	276	-2.1	-1.9	-9.1	-11.0	-6.3	-25.6
Brazil	494	887	798	809	4.8	-3.8	-1.3	-5.3	2.0	-8.3
Colombia	598	824	804	812	1.9	0.1	-1.2	-1.4	1.0	-1.5
Costa Rica	740	974	834	853	-2.1	-4.9	-9.7	-0.3	2.3	-12.4
Chile	958	1,015	895	935	6.2	4.1	-15.7	-2.4	4.2	-10.5
Ecuador	413	723	678	678	1.9	1.0	-1.1	-6.1	0.1	-6.2
El Salvador	422	433	344	339	-11.3	-10.9	-8.3	-2.9	-1.4	-21.8
Guatemala	418	589	512	497	0.9	-2.1	-6.2	-5.4	-2.6	-15.4
Haiti	90	114	99	100	5.1	-5.2	-4.9	-1.2	-0.1	-10.8
Honduras	313	356	318	314	-0.8	-2.3	-5.1	-3.8	0.6	-11.3
Mexico	978	1,366	1,284	1,295	5.5	5.1	-3.1	-7.7	0.9	-5.2
Nicaragua	418	337	331	317	6.7	2.0	-4.4	0.5	-4.7	-6.0
Panama	904	1,174	1,214	1,178	10.5	1.9	3.2	-1.8	-2.9	0.3
Paraguay	383	642	612	611	7.9	5.4	-3.9	-5.9	-0.9	-4.8
Peru	659	690	593	598	1.2	1.2	-2.2	-13.2	0.9	-13.3
Dominican Republic	398	601	615	605	3.6	1.6	-0.7	1.5	-1.7	0.7
Uruguay	1,097	1,426	1,226	1,195	5.3	1.2	-10.3	-5.3	-3.5	-16.2
Venezuela	1,239	1,310	1,147	1,097	-5.1	-3.3	-2.2	-7.4	-4.4	-16.2
Total	709	982	893	901	3.1	-0.7	-3.3	-5.3	0.9	-8.6

a. At market prices.
b. Provisional estimates subject to revision.
Source: ECLAC, on the basis of official figures.

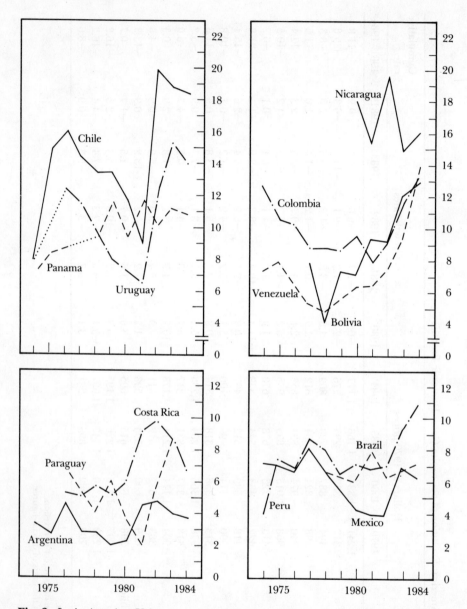

Fig. 2. Latin America: Urban unemployment, average annual rates (after ECLAC, on the basis of official information).

Brazil and Uruguay and declined 16 per cent in Peru and 11 per cent in Chile.

Notwithstanding the slowdown in economic activity, the increase in unemployment and the fall in real wages, as well as the weakening of external inflationary pressures, the rate of price increase accelerated in most countries during the crisis and in the region as a whole reached unprecedented levels in 1982 and again in 1983. Indeed, the simple average rate of increase of consumer prices rose from 29 per cent in 1981 to 47 per cent in 1982 and to 66 per cent in 1983, while the average rate weighted by the population more than doubled from 58 per cent in 1981 to 130 per cent in 1983 (see table 4). In fact, because the high rates of inflation occurred primarily in the countries with the largest population, in 1982 and again in 1983 two out of every three Latin Americans had to face increases in the price level of over 100 per cent per year.

These unfavourable changes in output, employment, wages, and prices were accompanied by and closely linked with other no less negative developments in the external sector. Between 1981 and 1983 the value of exports declined by almost 10 per cent, while imports had to be reduced by nearly one half in view of the virtual collapse of net capital inflow after the South Atlantic war between Argentina and the United Kingdom and the Mexican payments crisis of August 1982. Moreover, the steep decline in external financing and the simultaneous increase in factor payments brought about an abrupt reversal in the direction of resource transfers – with Latin America becoming in 1982 a net exporter of resources for the first time since 1968 – and contributed also to produce three consecutive balance of payments deficits in 1981–1983 (table 1).

Causes

As normally happens with complex economic phenomena, both external and internal factors played an important role in the gestation and precipitation of the crisis. Of course, the specific nature and relative importance of these factors were very different in the various national experiences. Thus, in some of them – and especially in certain Central American countries – the effects of social and political upheavals and the long and painful civil strifes which have marked their recent history were especially decisive, while in other cases mistaken economic policies played a more crucial role. Nevertheless, in most countries two factors were particularly important. The first was the excessive reliance on what can broadly be called as strategy of "debt-led growth"; the second was the abrupt and protracted deterioration in the external conditions that Latin American countries had to face both in trade and finance from 1980 onwards.

The Excesses of External Indebtedness Policy

In essence, the strategy of relying heavily on the use of foreign financing, which many Latin American countries adopted in the second half of the 1970s, was the domestic counterpart of two external factors. One was the situation of abundant

Table 4. Latin America: Evolution of consumer prices (variations from December to December)

Country	1976	1977	1978	1979	1980	1981	1982	1983	1984
Latin America[a]	62.2	40.0	39.0	54.1	56.1	57.6	84.6	130.9	184.2
Simple arithmetical average	35.5	23.6	25.4	35.8	27.3	28.8	47.5	66.2	162.6
Countries with traditionally high inflation	74.5	47.1	45.7	61.9	65.4	67.6	101.2	157.2	223.0
Argentina	347.5	150.4	169.8	139.7	87.6	131.2	209.7	433.7	683.4
Bolivia	5.5	10.5	13.5	45.5	23.9	25.2	296.5	328.5	2,177.2
Brazil	44.8	43.1	38.1	76.0	95.3	91.2	97.9	179.2	203.3
Colombia	25.9	29.3	17.8	29.8	26.5	27.5	24.1	16.5	18.3
Chile	174.3	63.5	30.3	38.9	31.2	9.5	20.7	23.6	23.0
Mexico	27.2	20.7	16.2	20.0	29.8	28.7	98.8	80.8	59.2
Peru	44.7	32.4	73.7	66.7	59.7	72.7	72.9	125.1	111.5
Uruguay	39.9	57.3	46.0	83.1	42.8	29.4	20.5	51.5	66.1
Countries with traditionally moderate inflation	7.9	8.8	9.8	20.1	15.4	14.1	12.2	16.4	18.5
Barbados	3.9	9.9	11.3	16.8	16.1	12.3	6.9	5.5	5.2
Costa Rica	4.4	5.3	8.1	13.2	17.8	65.1	81.7	10.7	17.4
Ecuador	13.1	9.8	11.8	9.0	14.5	17.9	24.3	52.5	25.1
El Salvador	5.2	14.9	14.6	14.8	18.6	11.6	13.8	15.5	9.8
Guatemala	18.9	7.4	9.1	13.7	9.1	8.7	−2.0	8.4	5.2
Guyana	9.2	9.0	20.0	19.4	8.5	29.1	—	—	—
Haiti	−1.4	5.5	5.5	15.4	15.3	16.4	6.2	12.2	6.1
Honduras	5.6	7.7	5.4	18.9	15.0	9.2	9.4	8.6	3.2
Jamaica	8.3	14.1	49.4	19.8	28.6	4.8	7.0	14.5	31.3
Nicaragua	6.2	10.2	4.3	70.3	24.8	23.2	22.2	32.9	50.2
Panama	4.8	4.8	5.0	10.0	14.4	4.8	3.7	2.0	0.9
Paraguay	3.4	9.4	16.8	35.7	8.9	15.0	4.2	14.1	29.8
Dominican Republic	7.0	8.5	1.8	26.2	4.2	7.4	7.1	9.8	38.2
Trinidad and Tobago	12.0	11.4	8.8	19.5	16.6	11.6	10.8	15.4	13.0
Venezuela	6.9	8.1	7.1	20.5	19.6	10.8	7.9	7.0	18.3

a. Totals for Latin America and partial figures for groups of countries represent average variations by countries, weighted by the population in each year.
Source: International Monetary Fund, *International Financial Statistics*, April 1985, and official information supplied by the countries.

international liquidity which prevailed between 1974 and 1981 and the other involved the extremely expansive policy followed during that period by the private international banks in their relations with many semi-industrialized countries.

The underlying foundation and justification of this strategy were the simple and not exactly novel idea of supplementing domestic savings with external resources in order to increase investment and hence the rate of economic expansion. However, the attraction and practical significance of this way of financing the growth process acquired radically different dimensions from the mid-1970s onwards, when there was an enormous increase in the volume of external resources being recycled in international capital markets after the first hike in oil prices and, in particular, when international interest rates turned negative in the mid-1970s due to the acceleration of inflation in the industrialized countries.

The possibilities opened up by this new international financial context were initially used by many Latin American countries in order to relieve the adverse effects which the recession in the industrialized economies and the sudden and substantial rise in the international price of petroleum had had on the purchasing power of their exports. Thus, thanks partly to the procurement of a greater volume of external resources, adjustment to the 1973 oil price rise was gradual and hence expansive. In fact, not only did Latin America maintain in 1974 a high rate of growth of close to 7 per cent but it also managed to increase its domestic product by nearly 4 per cent in 1975. Thus the region succeeded in riding out the post-1973 oil crisis with only a slowdown in the rate of growth, unlike the OECD economies whose output declined in absolute terms.

In the following years, and in spite of the strong and sustained expansion of their exports, the majority of the economies of Latin America continued to make intensive use of the abundant financial resources offered by the private international banks. In this way, the persistent accentuation of external indebtedness became both a salient characteristic and a basic requirement of their development processes.

The heavy reliance on external financing was, however, a double-edged sword. On the one hand, it made it possible the financing of higher levels of imports and capital formation, thereby helping to maintain rates of economic growth in many countries that were higher than would otherwise have been possible. On the other hand, it helped to maintain economic policies that were bound to increase inflationary pressures and/or lead eventually to balance-of-payments crises. Thus, in some countries excessive foreign borrowing enabled governments to expand public expenditure at very high rates while at the same time to repress inflation by keeping artificially low the prices of basic consumer goods and public utilities through generous subsidies. In other countries the plentiful supply of external loans made it possible to maintain during several years exchange-rate policies whose central aim was not to keep external equilibrium but to reduce inflation through its effects on expectations and the limits imposed on domestic prices by the fixing of exchange rates and the simultaneous and indiscriminate liberalization of imports.

As was to be expected, in both the countries that overexpanded domestic expenditure and in those which used exchange-rate policy as a fundamental tool of stabilization programmes, a common result was a spectacular increase in imports and a persistent loss of competitiveness of both export and import-substituting activities. Nevertheless, in spite of the very fast rise in the trade deficit that these changes brought about, no opportune corrective measures were taken since the ever-increasing flow of external loans made it possible not only to finance the import surplus and the rapidly rising flow of interest payments but also to build up international reserves.

Under these circumstances, the policies leading to an excessive growth of internal demand and of setting artificially low real exchange rates were maintained much longer than would have been possible had the supply of external financing been less abundant. But this also meant that the accumulation of both internal and external disequilibria was correspondingly larger. Hence, notwithstanding its relatively high economic growth rate during the second half of the 1970s – which far exceeded that of the OECD countries – the region was still very vulnerable to negative changes in the international environment.

This vulnerability became apparent when Latin America began to experience the full impact of the protracted recession that started in the industrialized economies in 1980 and had to face the substantial changes which occurred more or less simultaneously in the international capital markets.

These events affected the development of the region in three main ways. The first and most traditional was the deterioration of the terms of trade; the second was the dramatic rise in the real level of international interest rates; and the third – and most devastating – was the sharp drop in the net inflow of capital.

The International Recession and the Deterioration of the Terms of Trade

As had been the case during other recessions, the fall of economic activity in the industrialized countries diminished their demand for imports and contributed to reduce the rate of growth of international trade. This time, however, these negative consequences were aggravated by the revival of protectionist practices in many of the central economies, which became more frequent and stringent as unemployment grew and the recession continued. Under such circumstances, the volume of international trade, whose accelerated growth had played a fundamental role in the expansion of the world economy during the post-war period, rose very little in 1980, stagnated in 1981, fell by 2 per cent in 1982, and experienced only a weak recovery in 1983.

This evolution of international trade had extremely harsh consequences on Latin America, as it brought about a sharp drop in the international prices of most commodities. As a result of this, Latin America's terms of trade fell steadily during 1981–1983, accumulating a loss of about 23 per cent. The consequences of this trend were particularly serious in the case of the non-oil-exporting countries, whose terms of trade had already declined sharply during the triennium 1978–1980. Thus, between 1977 and 1983, the terms of trade of the region's non-oil-exporting countries fell by almost 38 per cent. So their average level for the period 1981–1983 was considerably lower than during 1931–1933, the most critical period of the Great Depression.

The Rise in International Interest Rates

A second external change that contributed to the crisis was the steep rise in international interest rates, starting in 1978, which in the industrial economies brought real rates of interest to their highest levels in almost half a century.

These exceptionally high interest rates contributed towards the unleashing and aggravation of the crisis of Latin America in two main ways. First, since they coincided with the decline in the region's terms of trade, they brought about an increase in the real cost of external credit for the Latin American countries that far exceeded the rise of real interest rates faced by borrowers in the industrial economies (see figure 3).

Second, because the proportion of total debt contracted at variable interest rates had increased very fast in most Latin American countries during the 1970s, the rise in interest rates led to an enormous growth in interest remittances. In fact, the value of these soared from under 6.9 billion dollars in 1977 to about 36 billion in 1983, thus growing over this period in a proportion of 415 per cent that more than doubled the 195 per cent expansion of the total external debt.

The increase in interest payments also surpassed by a large margin that of the total value of goods and services, especially between 1980 and 1982. In fact, the ratio of interest payments to exports doubled in just two years, reaching nearly 40 per cent in 1982 for the region as a whole and 47 per cent for the non-oil-exporting countries.

The Decline of Net Capital Inflow and the Negative Transfer of Resources

All things considered, however, what most contributed to precipitate the crisis was the sharp drop in the net inflow of capital, which began in 1982 and continued in 1983.

The negative impact of this decline in the inflow of loans and investments was particularly severe for three reasons. The first was that it took place after a long period during which the net inflow of capital had risen considerably and during which not only the equilibrium of the external sector but also the overall operation of many Latin American economies had come to depend considerably on the constant increase in external financing.

In the second place, the fall in the supply of external loans was procyclical in nature, as it occurred simultaneously with the decline in the purchasing power

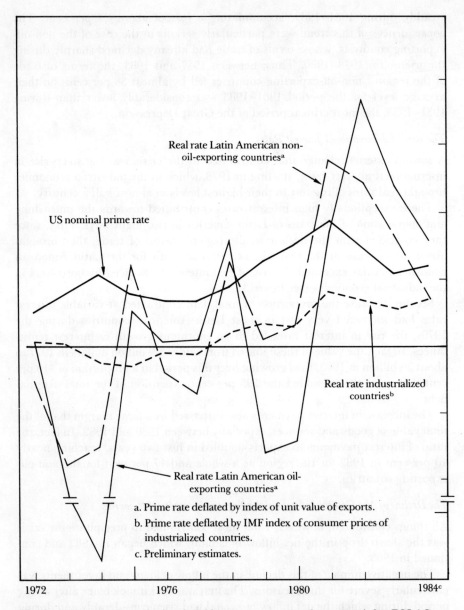

Fig. 3. International nominal and real interest rates, annual averages (after ECLAC, on the basis of official data, and the Economic Report of the President, February 1985).

of exports and the severe deterioration in the terms of trade caused by the international recession.

Finally, the decrease in the net inflow of capital was of an unusual magnitude and this, along with the simultaneous increase in net interest and profit remittances, led to a dramatic reversal in the direction of the transfer of resources between Latin America and the industrialized world.

The exceptional dimension of these changes is clearly brought out by the figures in table 5, which show how the net inflow of capital, after increasing steadily during the past decade and reaching a record figure of almost 38 billion dollars in 1981, plunged to just over 19 billion in 1982 and to a mere 4.4 billion in 1983.

Such a radical drop in external financing would have been dangerous under any circumstances. In this instance, however, its negative effects were compounded by the simultaneous and also sizeable increase in factor payments. Because of these changes, the balance between the two financial flows was drastically altered. As figure 4 shows, up to 1981 the net amount of foreign loans and investment was well over the amount paid out for interest and profits. Beginning in 1982, however, this situation was reversed as factor payments continued to increase and the net inflow of capital was nearly halved, thereby forcing Latin America to transfer to the rest of the world a considerable amount of real resources. The situation continued to worsen in 1983, when the slight fall in interest and profit remittances caused by the temporary decline of nominal interest rates was more than offset by a new and very large contraction in net capital inflows.

Moreover, this negative transfer of resources was extremely large, amounting to almost 50 billion dollars during the 1982-1983 biennium, a sum equivalent to about one quarter of the value of the region's exports of goods and services in the same period. Since the net resources received by Latin America in 1981 had represented the equivalent of almost 9 per cent of exports, the net result of the shift in the direction of the transfer of resources was equivalent to a deterioration in the terms of trade of nearly 40 per cent in just two years, an effect which more than doubled the actual fall of the terms of trade in 1982-1983.

To conclude, the basic cause of the severity and duration of Latin America's recent economic crisis was the unusual combination of three extremely unfavourable external events: a protracted world recession, abnormally high international interest rates, and the decision of the international private banks to cut back lending sharply as of 1982. Primarily, because of the combined effect of these changes, the capacity to import experienced a dramatic decline between 1981 and 1983, which far exceeded the reduction of exports (fig. 4). That peculiar combination of unfavourable external events also forced most countries to initiate at the same time a process of internal adjustment and negotiations with the international commercial banks aimed at refinancing and restructuring the service of the external debt.

Table 5. Latin America: Net inflow of capital and net transfer of resources (billions of dollars)

Year	New inflow of capital (1)	Net payments of profits and interest (2)	Transfer of net resources (=(1)−(2)) (3)	Net transfer of resources in real terms[a] (4)	Exports of goods and services (5)	Net transfer of resources/ export of goods and services (=(3)/(5))[b] (6)	Changes in the terms of trade (7)
1973	7.8	4.2	3.6	8.3	28.9	12.5	14.1
1974	11.4	5.0	6.4	12.5	43.6	14.7	15.9
1975	14.2	5.5	8.7	15.5	41.1	21.2	−13.5
1976	18.2	6.8	11.4	19.4	47.3	24.1	4.4
1977	17.0	8.2	8.8	14.1	55.9	15.7	6.0
1978	26.1	10.2	15.9	23.7	61.4	25.9	−10.5
1979	28.6	13.6	15.0	19.8	82.0	18.3	4.2
1980	30.0	18.0	12.0	13.9	105.8	11.3	4.1
1981	37.7	27.7	10.0	10.6	114.1	8.8	−9.2
1982	19.2	37.6	−18.4	−19.1	101.9	−18.1	−8.9
1983	4.4	34.5	−30.1	−30.9	100.5	−30.0	−6.2
1984[c]	12.4	37.4	−25.0	−25.0	113.0	−22.1	0.2

a. Obtained by deflating column 3 by the United States wholesale price index, base 1984 = 100.
b. Percentage.
c. Provisional estimates subject to revision.
Source: International Monetary Fund, *Balance of Payments Yearbook* (several issues); and ECLAC estimates, on the basis of official figures.

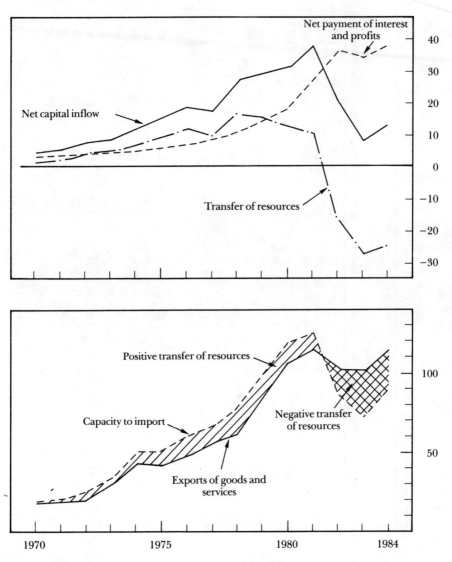

Fig. 4. Latin America: Net capital inflow, transfer of resources, and capacity to import (billions of dollars).

Adjustment (1982–1984)

Because of the abrupt fall in net capital inflow Latin America could no longer finance current account deficits of the colossal magnitude it had run in 1981–1982, amounting to over 35 per cent of total exports for the whole region and

Table 6. Indices of real effective exchange rates (1980 = 100)

Annual and quarterly averages[a]	Argentina	Bolivia	Brazil	Colombia	Costa Rica	Chile	Ecuador	Mexico	Peru	Uruguay	Venezuela
1975	192	97	81	118	107	136	112	104	77	130	117
1976	126	103	78	112	104	117	105	109	81	137	115
1977	164	99	78	102	105	114	99	121	90	136	107
1978	148	103	83	103	107	133	96	114	113	134	107
1979	112	105	93	99	103	119	99	109	107	110	108
1980	100	100	100	100	100	100	100	100	100	100	100
1981	131	70	83	92	143	89	96	90	87	94	91
1982	160	115	80	86	128	105	97	132	86	102	85
1983	155	77	95	87	112	112	102	135	94	138	79
1984	145[b]	56	89	89	—	111	116	114[b]	95	126	112[c]
1982											
1	138	94	84	89	152	93	92	105	83	95	87
2	147	119	79	87	131	96	95	124	84	96	86
3	187	121	77	85	118	111	104	157	86	91	84
4	159	117	81	85	115	116	98	137	88	124	84
1983											
1	158	108	93	87	111	123	98	147	90	146	82
2	164	83	101	85	109	112	106	135	94	139	80
3	149	59	95	87	112	105	102	132	96	135	78
4	149	60	91	90	113	109	103	131	95	132	77

1984							
1	154	54	91	90	114	111	121	91	130	97
2	136	76	92	88	112[d]	110	115	114	94	133	124
3	135	56	88	88	111[d]	105	120	111	95	119	115
4	159[e]	39	84	90	—	114	119	107[e]	99	120	—

a. Export trade-weighted average of the real exchange rate of the country with its major trading partners. From 1975 to 1979 the weights are yearly averages; from 1980 the 1980–83 average weight is used. For more information see the Technical Appendix of the *Economic Survey of Latin America*, 1981.
b. Average January–November.
c. Average January–September.
d. Preliminary estimation.
e. Average October–November.
Source: CEPAL, based on official information from the countries and the IMF's *International Financial Statistics*.

more than 50 per cent in the case of the non-oil-exporting countries. Adjustment policies to reduce external disequilibrium hence became mandatory in nearly all countries. In addition, in those of them where inflation had accelerated sharply, stabilization policies had to be applied as well.

In nearly all cases adjustment programmes were carried out within the framework of standby agreements with the International Monetary Fund. The programmes included varying combinations of domestic expenditure-reducing policies – implemented through restrictive fiscal, monetary and income policies – and of switching policies aimed at raising the relative prices of tradables by means of devaluations, higher tariffs, and export promotion measures.

Naturally, there were substantial differences in how strictly and persistently these policies were applied in the various countries. There was also considerable diversity in the extent to which adjustment and stabilization policies reached their respective objectives, though in general the latter were much less successful in controlling inflation than the former were in reducing external disequilibrium.

In fact, the rapidity with which the current account deficit was closed was amazing and surpassed the most optimistic expectations and forecasts. In merely two years the region's current account deficit plummeted from 40 billion dollars in 1982 to slightly over 2 billion dollars in 1984. This dramatic change was mirrored by the equally spectacular turnaround experienced by the trade account, which moved from a deficit of 2 billion dollars in 1981 to a surplus of over 38 billion in 1984.

Nevertheless, because of the extremely short period in which it was achieved and the circumstances prevailing in the world economy, this sharp reduction of the current account deficit entailed high costs in terms of output and employment. In effect, because of the time constraint, the adjustment process was basically incompatible with the real reallocation of resources from the production of non-tradables to the production of import-substitutes and especially of exports, a process which is basic for positive, non-recessionary adjustment but which can only be carried out over an extended period of time. Moreover, until 1983 the growth of exports was limited by the drop in the international price of most of the main commodities exported by the region and by the initial decline and subsequent sluggish recovery of world trade. Thus, despite the substantial increases in real effective exchange rates that, as shown in table 6, resulted from the devaluations implemented in many countries since 1981, the value of exports fell sharply in 1982 and stagnated in 1983.

Consequently, the spectacular turnaround of about 33 billion dollars in the trade balance between 1981 and 1983 was due entirely to the contraction of imports. As may be seen in table 7, the value of imports declined precipitously from almost 96 billion dollars in 1981 to less than 55 billion in 1983, while their volume fell by an almost incredible 41 per cent in just those two years. The quantum of imports plunged even more sharply in Argentina and Chile (where

Table 7. Latin America: Imports of goods

	Value (millions of dollars)			Quantum (growth rates)				
	1981	1983	1984[a]	1982	1983	1984[a]	1982–1983[b]	1982[a]–1982[b]
Latin America	95,809	54,742	57,000	−18.7	−27.2	4.7	−40.8	−37.8
Oil-exporting countries	43,005	19,117	21,830	−20.1	−41.9	12.4	−53.6	−47.7
Bolivia	680	482	470	−35.3	16.2	−9.9	−24.8	−31.4
Ecuador	2,362	1,408	1,580	−7.6	−34.8	12.2	−39.8	−32.3
Mexico	24,038	7,723	10,000	−43.4	−42.3	26.1	−67.3	−58.7
Peru	3,802	2,723	2,430	−4.1	−29.2	−14.2	−32.1	−41.7
Venezuela	12,123	6,781	7,350	18.8	−47.2	8.2	−37.3	−32.2
Non-oil-exporting countries	52,804	35,625	35,170	−17.1	−11.1	−0.7	−26.3	−26.8
Argentina	8,432	4,121	4,270	−39.4	−11.3	1.5	−46.3	−45.4
Brazil	22,091	15,429	13,940	−8.9	−15.3	−4.6	−22.8	−26.4
Colombia	4,763	4,759	4,720	18.7	−6.1	−3.0	11.5	8.2
Costa Rica	1,090	893	1,000	−30.1	7.8	18.8	−24.7	−10.4
Chile	6,513	2,838	3,360	−39.4	−15.7	16.0	−48.9	−40.7
El Salvador	898	803	870	−13.0	−4.4	9.2	−16.9	−8.9
Guatemala	1,540	1,056	1,150	−21.6	−20.3	3.2	−37.5	−35.3
Haiti	398	352	410	−21.2	2.0	−6.6	−19.7	−24.0
Honduras	899	761	840	−25.0	12.9	−2.3	−15.4	−17.4
Nicaragua	923	778	780	−25.5	3.7	−1.2	−22.8	−23.8
Panama	1,441	1,246	1,240	−4.6	−14.9	−14.9	−18.4	−31.0
Paraguay	772	552	580	−11.2	−19.1	1.1	−28.1	−27.4
Dominican Republic	1,452	1,297	1,220	−15.3	4.9	−6.6	−11.1	−16.6
Uruguay	1,592	740	790	−29.8	−22.2	8.1	−45.4	−40.6

a. Provisional estimates.
b. Cumulative rates.
Source: ECLAC, on the basis of official figures.

it was halved between 1981 and 1983), in Venezuela (where the volume of imports fell by nearly 50 per cent in 1983 alone), in Uruguay (which reduced its real imports by over 55 per cent during the triennium 1981–1983) and, above all, in Mexico (which brought down its imports by more than 67 per cent from 1981 to 1983).

Although these colossal reductions reflected in part the abnormally high levels imports had reached in 1981, they hit not only the purchases of luxury or expandable consumer goods but also cut deeply into the imports of machinery and indispensable intermediate inputs. It was not surprising, therefore, that in 1982 total output remained at a standstill or declined in many countries and that it fell again and more severely in 1983 in most of them.

Thus up to 1983 the region's adjustment, though speedy and successful in terms of closing the external disequilibrium, was anything but efficient from the points of view of output and employment. For there can generally be no efficient "shock" adjustment to external disequilibrium, since efficient adjustment, unlike stabilization policies, requires changes in the allocation of real productive resources, not simply in the nominal values of variables. For this reason gradualism is central to efficient adjustment. For not only must the output of non-tradables decline in relative terms, which can occur rapidly, but the output of tradables must rise, and this necessarily is a slower process; not only must the volume of imports be reduced, which also can occur rapidly, but the output of exports and import substitutes must increase, and this takes time.

The nature and sources of the adjustment process began to change, however, in 1984. Favoured by higher real exchange rates and other incentives, and taking advantage of the buoyancy of world trade, the value of merchandise exports rose 10 per cent, thus surpassing slightly its pre-crisis level.

This recovery of exports made it possible to finance a moderate increase of almost 5 per cent in the volume of imports, which, in turn, marginally diminished the critical bottleneck represented by the scarcity of foreign inputs and raw materials. Thus, in contrast to what had happened in the two preceding years – during which, as already noted, the bigger positive balance recorded in merchandise trade was entirely due to successive and very sharp reductions in the value of imports – in 1984 the increase in the trade surplus resulted exclusively from the expansion of exports and was hence compatible with the recovery of domestic economic activity.

Furthermore, since the growth of the trade surplus more than offset the rise in net payments of profits and interest, the current account deficit diminished for the second year in a row and fell to a level of 2.1 billion dollars, equivalent to barely 5 per cent of the huge negative balance recorded just two years earlier.

Recovery (1984–?)

In general, 1984 represented in Latin America a clear improvement over the dismal performance of the economy during the preceding three years, though, as will be explained later, it would be premature to consider that it also represented a clear turning-point.

In the external sector, in addition to the virtual elimination of the current account deficit, there was a moderate revival in the net inflow of capital, reversing the pronounced downward trend that had characterized the two preceding years. Thanks to this increase, and also because of the lower deficit on current account, the balance of payments closed with a surplus of about 10 billion dollars. This positive balance – the first since 1980 and the largest ever recorded in Latin America – permitted a partial recovery in the level of international reserves, which had dropped by almost 30 billion dollars in the three preceding years.

Furthermore, in 1984 the growth rate of Latin America's total external debt continued to slacken. According to provisional estimates, the stock of disbursed external debt amounted to about 360 billion dollars at the end of the year. Thus it grew by 5.5 per cent, i.e. less than the 8 per cent increase of 1983 and far below the 14 and 24 per cent increases recorded in 1982 and 1981. Since the growth rate of the external debt was likewise slower than that of exports, the debt export coefficient decreased for the first time in the last four years.

During 1984 there was also an interruption of the sharp downward trend in the rate of economic growth which had been occurring since 1979. In fact, after falling 1 per cent in 1982 and 3 per cent in 1983, the region's gross domestic product went up by 3.3 per cent. Moreover, the change was widespread, with GDP rising in 14 of the 19 countries for which comparable data are available (table 2).

In the region as a whole, however, the increase in economic activity barely exceeded that of the population. Per capita product therefore rose less than 1 per cent above the very low level to which it had sunk in 1983 and thus was equivalent only to that reached in 1977. Moreover, inasmuch as the recovery tended to be concentrated in the region's large and medium-sized economies, and was very weak in most of the smaller countries, per capita product once again diminished in 9 countries.

The inadequacy of the recovery was also evidenced by the fact that urban unemployment rates continued to climb in Brazil, Nicaragua, Bolivia, Colombia, Peru, and Venezuela – reaching record levels in the last four countries – and declined only slightly in Argentina, Mexico, Chile, and Uruguay – in the last two cases from the extremely high levels recorded in 1983 (fig. 2).

Hence, in most countries of the region the 1984 recovery was weak and insufficient, and in a good number of them simply did not take place. Moreover, the possibility of applying expansionary policies leading to a resumption of economic growth continued to be limited in many countries by the unprecedented levels reached by the inflationary process.

In effect, in spite of the frailty of the economic recovery in most countries and the attenuation of external inflationary pressures, the rate of increase of prices rose in half the Latin American economies and in the region as a whole reached a new all-time high, with the simple average rate of inflation shooting up from 66 per cent in 1983 to over 160 per cent in 1984, and the rate weighted by the population jumping in the same period from 130 to 184 per cent.

In particular, inflation gathered enormous speed in 1984 in Argentina (680 per cent) and especially in Bolivia (2,200 per cent), exceeded 200 per cent in Brazil, remained above 100 per cent in Peru, declined only from 80 to 60 per cent in Mexico, and accelerated sharply, although starting from very different levels, in Uruguay, Nicaragua, Paraguay, Jamaica, the Dominican Republic, and Venezuela (table 4).

This upward trend persisted in many countries in 1985. Thus, during just the first quarter consumer prices shot up by 500 per cent in Bolivia, nearly doubled in Argentina, and climbed 40 per cent in Brazil and 35 per cent in Peru. As a result, in these four countries the yearly rates of inflation are running now at the unprecedented levels of 8,200, 850, 225, and 130 per cent respectively. Moreover, during the first quarter the yearly rate of increase of consumer prices continued to oscillate around 65 per cent in Uruguay, tended to stabilize at just under 60 per cent in Mexico and rose steadily in both Chile and Colombia (see figure 5).

High inflation is not, however, the only element presently blocking the way to recovery and the resumption of sustained growth. An equally important obstacle is the continued and precarious balance-of-payments situation in most countries of the region.

At first glance, this may seem surprising in view of the substantial improvement which, as already noted, was achieved in the external sector in 1984, and considering also the widespread belief in certain influential circles of the international financial community that the Latin American "debt crisis" is a thing of the past and that by now what exists is, at worst, a Latin American "debt problem."

Nevertheless, it should be remembered, in the first place, that the sizeable reduction of the region's current account deficit in 1984 was due primarily to the 23 per cent expansion of exports of Brazil and the large current account surpluses obtained by Mexico and Venezuela. In all the other countries of the region, the current account closed with deficits, which, in several cases, were actually larger than those recorded in 1983. Thus, in the absence of significant falls in international interest rates or of sudden and sharp improvements in their terms of trade, many countries will need to continue to implement restrictive domestic policies in order to reduce their external disequilibria.

A second element pertaining to the external sector that restricts the possibilities for increasing the pace of economic growth is the still extremely low levels of both total imports and import coefficients in practically all countries. Notwithstanding its recent recovery, the volume of imports of the whole region was still 38 per cent lower in 1984 than in 1981. Moreover, this was a very widespread characteristic, Colombia being the only country in which the quantum of im-

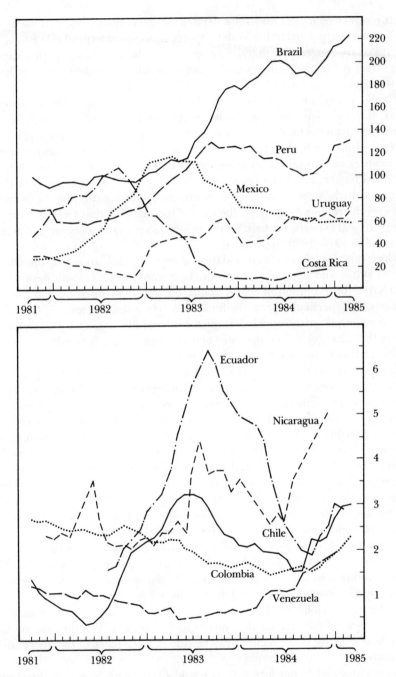

Fig. 5. Latin America: Twelve monthly variations in the consumer price index (percentages).

ports rose between 1981 and 1984. Under these circumstances, and owing also to the even sharper reduction in the imports of luxury or expendable goods, any expansionary programme that did not go hand in hand with a growth of exports or a larger net capital inflow would quickly be limited by the foreign exchange bottleneck.

Finally, it ought to be noticed that, despite the improvement shown in recent years, the coefficients relating to the external debt are still very high in most Latin American countries, both in international comparative terms and in comparison with the situation that prevailed before the crisis. For example, the 3.3 debt-export ratio recorded in 1984, even though lower than that of 1983, was one-third higher than those recorded on average in 1978–1980, and this difference was much larger in the case of the non-oil-exporting countries. The change was even more unfavourable in the case of the ratio relating total interest payments to total exports, as the 1984 coefficient (35 per cent) doubled the average recorded in 1977–1980 (see table 8).

Thus, it is clear that the consolidation of recovery and, above all, the reinitiation of stable and dynamic growth in the immediate future will depend very closely on the evolution of world trade, on trends in international financial markets, and, particularly, on the future levels of interest rates.

A few simple arithmetical examples can usefully illustrate the significant effects that changes in interest rates and the terms of trade would have on the region's capacity to import and, hence, to grow.

Since the region's debt-export ratio is about 3.3, since around 70 per cent of the debt has been contracted at variable interest rates, and since exports are presently 40 per cent larger than imports, every percentage point of reduction (increase) in international interest rates would permit (force) imports to be increased (reduced) by about 3.2 per cent without altering the resultant current account. Thus, if current international interest rates and/or spreads fell by three percentage points – which would still leave them above their historic rates in real terms – import volumes could rise by nearly 10 per cent, thus permitting faster growth, without this bringing about any loss of international reserves.

Similarly, if the export prices of the non-oil-exporting countries of the region recovered their average levels of 1950–1970 and the real price of oil remained at its present level, the terms of trade of these countries would improve by 30 per cent, allowing them to finance 70 per cent of interest payments, or increase their import volumes by 40 per cent (the same 40 per cent they sacrificed in 1982–1984), or run down their nominal debt by over 7 per cent per year.

To be sure, such figures are averages and they obviously vary from country to country. Nor, obviously, are they meant to be forecasts. But they clearly illustrate how sensitive is the region's future growth to changes in the international economy.

In any case, while not forecasts, it would perhaps be an unduly pessimistic scenario which presumed that in the coming years the region's terms of trade

Table 8. Latin America: Ratio of total interest payments to exports of goods and services (percentages)[a]

Country	1977	1978	1979	1980	1981	1982	1983	1984[b]
Latin America	12.4	15.5	17.4	19.9	26.4	39.0	35.8	35.0
Oil-exporting countries	13.0	16.0	15.7	16.5	22.3	32.0	31.0	33.0
Bolivia	9.9	13.7	18.1	24.5	35.5	43.6	49.3	57.0
Ecuador	4.8	10.3	13.6	18.2	24.3	30.1	26.0	31.5
Mexico	25.4	24.0	24.8	23.1	28.7	39.9	36.7	36.5
Peru	17.9	21.2	14.7	16.0	21.8	24.7	31.2	35.5
Venezuela	4.0	7.2	6.9	8.1	12.7	21.0	20.3	25.0
Non-oil-exporting countries	11.9	15.1	18.8	23.3	31.3	46.6	40.7	36.5
Argentina	7.6	9.6	12.8	22.0	31.7	54.6	58.4	52.0
Brazil	18.9	24.5	31.5	34.1	40.4	57.1	43.4	36.5
Colombia	7.4	7.7	10.1	13.3	21.6	25.0	21.7	21.5
Costa Rica	7.1	9.9	12.8	18.0	25.5	33.4	41.8	32.0
Chile	13.7	17.0	16.5	19.3	34.6	49.5	39.4	45.5
El Salvador	2.9	5.1	5.3	6.5	7.5	11.9	14.2	15.0
Guatemala	2.4	3.6	3.1	5.3	7.5	7.8	7.6	4.0
Haiti	2.3	2.8	3.3	2.0	3.2	2.4	4.9	5.0
Honduras	7.2	8.2	8.6	10.6	14.5	22.4	17.7	19.0
Nicaragua	7.0	9.3	9.7	15.7	15.5	33.2	19.3	18.5
Paraguay	6.7	8.5	10.7	14.3	15.9	14.9	24.3	19.0
Dominican Republic	8.8	14.0	14.4	14.7	10.5	22.6	24.9	23.5
Uruguay	9.8	10.4	9.0	11.0	13.1	22.4	27.6	31.5

a. Interest payments include those on the short-term debt.
b. Provisional estimates subject to revision.
Source: 1977–1983: International Monetary Fund, *Balance of Payments Yearbook*; 1984: ECLAC, on the basis of official data.

were more likely to worsen than improve, that international interest rates (plus spreads) were more likely to rise than fall, and that export volumes would fall rather than rise. All the more so when the fruits of the policies most countries of the region implemented between 1982 and 1984 – such as real devaluation and the correction of other distorted, key relative prices – have only partially emerged. For, as already noted, time is required for these policies fully to effect the required shift in productive resources. As these positive, longer-term effects take hold, the foreign exchange constraint should slowly ease, thus permitting the pursuit of more expansive demand management policies and the fuller use of existing productive capacity.

Needless to say, however, this possibility would be jeopardized if stagnation or recession occurred in the central countries, events which, obviously, Latin American economic authorities cannot influence and which, as past forecasting experience shows, can hardly be accurately predicted. In the end, therefore, the possibility of attaining sustained, more autonomous and rapid economic growth in the region, though conditioned by external factors, will depend more closely on the firm and stable implementation of correct and coherent domestic economic policies and on the rise of Latin Americans' propensities to work, save, and innovate. Seen from this angle, the better understanding and wider acceptance of these time-honoured principles of sound development by the public, and especially by a growing number of political leaders of widely different ideological persuasions, may prove to be the one and lasting positive consequence for the region of its recent traumatic economic crisis.

THE RISE AND FALL OF CAPITAL MARKETS IN THE SOUTHERN CONE

Joseph Ramos

Introduction

Three economic problems contributed to the political upheavals and military coups which gave rise to neo-conservative experiences in Argentina, Chile, and Uruguay in the mid-1970s: (1) galloping inflation (triple-digit); (2) severe balance of payments difficulties; and (3) slow growth. The first two problems, of a short-term nature, were met with price stabilization and adjustment policies of a fairly orthodox, monetarist bent. The latter was met by the widespread liberalization of the economic system, of which financial liberalization was but one part.

The radical liberalization and restructuring which the neo-conservatives effected was their response to the strikingly poor economic performance of these countries in the post-war period. Since 1945, the per capita income of the Southern Cone countries had grown by only 1.5 per cent per year, as compared to 3.4 per cent for the rest of Latin America (see table 1), so that by the mid-1970s their share in the region's GNP had fallen from over a third to less than a quarter. And even though Argentina still enjoyed the highest per capita income in Latin America, by the mid-1970s Chile had fallen from third to seventh place, and Uruguay from second to fifth. Such poor economic performance was attributed by neo-conservatives to the exaggerated and increasingly discretional intervention of the state, an outgrowth of the Great Depression and govern-

---

This paper was prepared during the author's stay at the Kellogg Institute for International Studies of the University of Notre Dame in January–February 1985.

Table 1. Southern Cone

Country	Historical rate	Neo-conservative period	Phase 1	Phase 2	Phase 3[a]
			Annual growth rate of per capita Gross Domestic Product		
Argentina	1950–1975: 1.7	1976–1983: −1.3	1976–1978: −0.9	1979–1980: 2.4	1981–1983: −4.3
Chile	1950–1973: 1.5	1974–1983: −0.6	1974–1976: −4.7	1977–1981: 6.2	1982–1983: −9.1
Uruguay	1950–1974: 0.6	1975–1983: 1.2	1975–1978: 3.6	1979–1980: 5.3	1981–1983: −4.8
Latin America (excludes Southern Cone)	1950–1974: 3.4	1975–1983: 1.0	1975–1978: 2.9	1979–1980: 3.8	1981–1983: −2.7
			Share of national savings in Gross National Product		
Argentina	1966–1975: 19.9	1976–1983: 18.9	1976–1978: 23.7	1979–1980: 20.5	1981–1983: 15.0
Chile	1964–1973: 17.0	1974–1983: 12.2	1974–1976: 16.4	1977–1981: 12.0	1982–1983: 3.8
Uruguay	1965–1974: 10.4	1975–1983: 13.2	1975–1978: 12.4	1979–1980: 16.0	1981–1983: 13.1
			Share of investment in Gross National Product		
Argentina	1966–1975: 20.0	1976–1983: 20.6	1976–1978: 22.2	1979–1980: 22.9	1981–1983: 17.4
Chile	1964–1973: 19.7	1974–1983: 17.3	1974–1976: 17.8	1977–1981: 19.7	1982–1983: 10.4
Uruguay	1965–1974: 10.2	1975–1983: 15.8	1975–1978: 14.5	1979–1980: 20.6	1981–1983: 14.5

a. 1983: preliminary figures.
Source: ECLAC, on the basis of official data and Central Bank of Chile.

ments' attempts to cope with it. The neo-conservatives proposed, therefore, to replace the interventionist paradigm which had prevailed since the Great Depression, and restore the market as the principal mechanism of resource allocation.

Since "financial repression" was one of the most characteristic elements of the interventionist period in the Southern Cone, it was among those most severely criticized by neo-conservative policy-makers from the start. For it was inconceivable for them that the central mechanism for determining the allocation of resources – the capital market – should be controlled not by the forces of supply and demand, but by the discretional authority of the government. For this reason, the creation of a domestic capital market and its opening up to the outside world made up part of their package of basic structural reforms right from the very beginning. Ironically, the end of the three neo-conservative experiences was accompanied, if not brought on, by the collapse of the very financial system they had created.

The Neo-conservative Diagnosis

Ever since the Great Depression, the countries of the Southern Cone had established increasing controls over the financial system. Selective credits at preferential interest rates – often negative in real terms – were created to promote the development of sectors and activities considered to be of the highest priority. Exchange controls were placed on capital movements in order to avoid capital flight and to render possible the maintenance of a low rate of exchange, and so cheapen the import of foodstuffs and intermediate inputs. Moreover, a goodly part of the banking system not only belonged to the state, but was administered in highly discretional form, so that credit was often assigned according to political rather than economic criteria.

In addition to their basic stance against intervention, neo-conservatives criticized this domestic financial repression for the following reasons:[1]

1. Low, or even negative, rates of interest were thought to explain why savings were so low in Chile (17 per cent) and Uruguay (10 per cent), or why they depended so heavily on the public sector in Argentina (see table 1). For such rates of interest provided little or no incentive for individuals to sacrifice current consumption.
2. Moreover, artificially low interest rates encouraged self-financing and discouraged financial intermediation. In this way, the market was segmented between those who had and those had no access to artificially cheap credit, all of which led to a poor allocation of resources. The former were induced to initiate projects with low rates of return, overmechanize or build in unnecessary capacity, whereas the expansion of capital-scarce activities with high rates of return was discouraged, forcing these activities to borrow

at the overblown interest rates of informal credit channels or condemning them to expand only to the limits of their capacity for self-financing. Such segmentation would indeed hurt capital-scarce activities with good investment opportunities, but, as the results will suggest, these were more likely to be small- and medium-size firms on the verge of modernizing, rather than, as some thinkers[2] favourable to financial liberalization believed, large firms already using modern technology.

3. The volume and variety of financial assets were severely limited in the economy. The proportion of M_2 in Southern Cone GNP at the beginning of the neo-conservative experiences (20 per cent) was well below that in the industrialized countries (60 per cent) or in some fast-developing underdeveloped countries (60 per cent in Taiwan, 33 per cent in the Republic of Korea and Mexico).[3] Moreover, the variety of financial instruments, especially for medium- and long-term debt, was quite limited.

The majority of economists were aware of these problems and concurred with the need to reduce the degree of financial repression. However, only the neo-conservatives *à outrance* thought that the solution was to leave the financial system wholly and entirely in the hands of the market. Those from other schools of thought believed that some form of control was indispensable, for the financial market is not like any other market. They argued that financial activity is intrinsically fragile, subject to abrupt and discontinuous changes (vicious and virtuous circles): once critical levels of confidence (or lack of confidence) regarding the future ability to service one's debts are reached, these tend to feed back and reinforce themselves. The liquidity problems of firms, for example, can lead to generalized insolvency if not attended to in timely fashion by the economic authorities. It was further pointed out that, given the relatively small size of the Southern Cone economies, plus the fact that there tend to be important economies of scale in financial activities, it was altogether likely that, if left to itself, the financial sector could quickly come to be controlled by a relatively few economic conglomerates with all the vices and defects that an oligopolistic allocation of credit entails.

As for capital inflows, neo-conservatives were one in insisting on the merits of an extensive financial opening up to the outside world as one of the key mechanisms by which a less developed country could take full advantage of the international economy, by fully utilizing foreign savings potential to augment domestic savings and so speed up growth rates. Nevertheless, neo-conservatives did differ among themselves as to the optimal sequence of liberalization in different markets. Some, such as McKinnon and Frenkel,[4] argued that first should come the opening up of trade and the creation of the domestic capital market, and only later, and gradually, financial opening up. They feared that should the financial opening up come early, interest rates would tend to converge before the prices of goods. Thus, investment would increase, but it would be misallocated inasmuch as relative prices would still be distorted. Others, such as Mundell, believed that such a risk was worth taking, believing that the heavy inflow of capital would in any case offset the initial contractionary effects of the devaluation, thus helping avoid a recession and so generating confidence in the overall liberalization process.

The Policies of Trade and Financial Opening Up[5]

Domestic capital markets were created in each of the three countries as of the first or second year of its neo-conservative experience (Chile in 1975, Uruguay in 1976, and Argentina in 1977). Moreover, Uruguay chose to accompany this with a wide financial opening up to the outside world but a timid trade opening up, whereas Chile liberalized trade first and only gradually opened up its capital account.

Chile's policy was conditioned by its very high inflation, and therefore its perceived need to assure control over the money supply. Moreover, the reluctance of the international banks to lend to Chile during the first years following the coup (for political as well as economic reasons) really made no other alternative possible. Argentina followed a middle road, controlling capital inflows during the first phase of its price stabilization programme (1976–1978) and increasingly opening itself financially during the second phase (mid-1978 onwards), once its stabilization efforts had centred on controlling the exchange rate.

The creation of domestic capital market included the following principle measures: (1) freeing interest rates; (2) eliminating or dramatically reducing existing qualitative and quantitative controls over credit (e.g. by sector of activity, type of collateral, size of firm, use of credit); (3) reducing the barriers to entry for new banks, financial intermediaries, and foreign banks (especially in Argentina and Chile); (4) the progressive reduction of reserve requirements; and (5) in Chile, the return or auction to the private sector of the bulk of the banks which had been placed under state control under Allende.

Financial opening up to the *outside* world included: (1) authorization to open banking accounts within the country denominated in foreign currency; and (2) the progressive reduction of limits on the entry and outflow of capital, both as regards the minimum time for such loans as well as limits on the amounts that could flow in. These limits were important in Argentina and Chile especially through 1978.

Argentina first prohibited the entry of foreign capital for periods of less than 180 days, and later raised it to one year (August 1977) and two years (November 1977). Moreover, borrowers were obliged to deposit the equivalent of 10–20 per cent of the foreign credit in domestic currency and at zero interest, all of which raised its effective cost to borrowers. These restrictions (which were only levied on private sector borrowing) were justified as measures necessary to maintain control over the growth of the money supply. However, once the price stabilization policy (in mid-1978) shifted from controlling the money supply to controlling the exchange rate, these restrictions were gradually eased. Indeed, to meet the heavy drain on reserves, even the inflow of very short-term capital was permitted as of mid-1980.

Chile, on the other hand, maintained the prohibition on capital inflows for periods under two years almost to the end (mid-1982). For it feared that unrestricted financial liberalization would bring in so much capital – given the huge differentials between domestic and international interest rates – that the stabilization programme could be jeopardized. In any case, Chile did tend to increase access in the course of time. At the beginning only non-financial enterprises could borrow; then banks were allowed to borrow up to certain limits; and finally those limits were substantially raised.

Even though the three countries established different sets of restrictions, capital inflows to the three did not differ all that much, for, in the final analysis, such inflows depend not only on the demand for credit (what the country wants and allows), but on its supply (what international banks are willing to lend under the conditions). Similarly, one of the really effective restrictions was the limits imposed on external borrowing by the public sector. Argentina and Uruguay increased the public sector's foreign indebtedness substantially right from the very beginning whereas Chile discouraged it almost to the very end.

The Principal Results of Financial Liberalization

The final objective of the policy of financial liberalization and opening up was to raise the level of domestic savings, increase investment, and improve resource allocation. The key policy instrument for this in the domestic plane was the freeing of interest rates. This was expected to encourage savings, equalize interest rates for all users (as between formal and informal credit segments) and lower the costs of financial intermediation, increasing the volume and variety of financial instruments. The freer flow of international capital was expected further to raise investment and move domestic interest rates closer to international ones.

The effects of liberalization and opening up were dramatic, but more often than not because they proved to be so different from what was expected. No doubt some of the unsatisfactory results were due not to financial liberalization itself but to unfavourable external conditions. Yet, as I will spell out shortly, a large part of the failures can be attributed to the questionable decision to pursue financial liberalization along with, rather than after, a price stabilization policy (both in its initial tight money variant and in its later variant of fixing or pre-announcing the exchange rate devaluation).[6] This error, along with others committed in the process of financial liberalization, grew out of the neo-conservatives' grossly simplified or mistaken assumptions as to the workings of the economy. The principal seven results, in summary form, were as follows.

First, as expected and desired, thanks to economic liberalization, financial intermediation strongly increased its share in GNP, rising by at least two percentage points (see table 2). More importantly, there was a remarkable increase in the proportion of GNP held in the form of time and savings deposits and of credit to the private sector (see table 3). These increased from threefold to over tenfold, as the case may be, between the beginning of the neo-conservative experience and the peak values achieved before its final crisis and demise.

Notwithstanding the wide variety of financial instruments generated by the liberalization of capital markets, the bulk of these were of very short-term duration (30 days and less). High interest rates on such short-term deposits, plus strong inflation and future uncertainty, made it very difficult subsequently to generate longer-run instruments which could be attractive to depositors as well as borrowers. Hence, the domestic capital market was never really anything other than a market in quasi-money. It was only toward the end of the process in phase 2 that significant long-term instruments were offered, and even then such interest rates ranged between 12 and 18 per cent real per year. But these never became any more than a small fraction of overall credit. And the market for long-term bonds was virtually non-existent.

Second, despite the remarkable increase in time and savings deposits, the proportion of GNP saved (that is, income not actually consumed),[7] was actually lower during the neo-conservative period than in the years immediately preceding that experience in both Argentina and Chile (table 1). Only Uruguay shows significant improvement in this regard. In short, whereas financial savings proved to be highly sensitive to interest rates, real domestic savings proved to be far less so.

Third, foreign savings (foreign debt) grew sharply during the neo-conservative period, partly as a response to highly favourable interest rates, but partly also to the generalized expansion of international liquidity in the period. In any case, by the end of 1983 the ratio of foreign debt to exports ranged from 3.3 to 1 in Uruguay and 3.8 to 1 in Chile to 4.9 to 1 in Argentina (see table 4), compared to an average of less than 3 to 1 for the rest of the region. What is surprising is not so much that foreign debt grew so strongly during the neo-conservative period (20 per cent per year) – after all it grew at a similar rate throughout the rest of Latin America – but that it grew so strongly when at the beginning of the neo-conservative experiences the countries of the Southern Cone were already among the most indebted countries of the region, at least in relation to the level of exports.[8]

It is notable that though Chile had the least internationally opened up financial sector and though its public sector was deliberately restrained from borrowing from abroad, it was Chile and not Uruguay that received the heaviest inflows of foreign capital throughout the neo-conservative period, not only in relation to GNP and exports but in absolute terms (see tables 4 and 5).[9]

Fourth, notwithstanding the sharp increase in capital inflows, and the noted increases in *financial* savings, investment as a proportion of GNP actually

Table 2. Southern Cone: Savings coefficients[a]

	Argentina				Chile				Uruguay			
	$\frac{GDS^b}{GNP^c}$	$\frac{GNS^d}{GNP}$	$\frac{FS^e}{GNP}$	$\frac{GNP_{fin}^f}{GNP}$	$\frac{GDS}{GNP}$	$\frac{GNS}{GNP}$	$\frac{FS}{GNP}$	$\frac{GNP_{fin}}{GNP}$	$\frac{GDS}{GNP}$	$\frac{GNS}{GNP}$	$\frac{FS}{GNP}$	$\frac{GNP_{fin}}{GNP}$
	(1)	(2)	(3)	(4)	(1)	(2)	(3)	(4)	(1)	(2)	(3)	(4)
1970	22.0	21.5	0.5	4.3	23.4	21.6	1.7	11.5	11.4	9.6	1.8	5.8
1971	25.5	24.3	1.2	4.3	20.8	17.8	2.9	11.1	11.2	8.9	2.3	5.9
1972	25.2	24.6	0.6	4.2	15.2	10.4	4.8	10.7	9.6	11.1	−1.5	6.0
1973	22.7	24.6	−1.9	4.1	14.3	9.5	4.8	10.1	9.1	9.6	−0.5	6.0
1974	22.4	22.7	−0.3	3.9	25.8	25.3	0.5	12.0	9.1	5.7	3.4	5.7
1975	22.1	19.4	2.7	4.3	14.0	8.5	5.6	13.8	10.9	6.5	4.4	5.5
1976	23.6	24.9	−1.3	3.9	13.6	15.4	−1.9	13.8	12.7	11.0	1.7	5.3
1977	26.2	23.9	2.3	4.3	14.4	10.7	3.7	13.4	14.8	11.6	3.2	5.2
1978	24.3	27.5	−3.2		16.5	11.6	4.8	13.7	16.0	13.8	2.2	9.5
1979	25.7	24.9	0.8	8.0	19.6	13.7	5.9	14.4	18.7	13.7	5.0	—
1980	26.6	22.9	4.4	8.0	23.9	17.9	6.0	15.2	20.1	14.5	5.6	—
1981	22.5	18.0	4.5	9.0	23.9	11.3	12.6	15.9	16.8	13.5	3.3	—
1982	19.0	16.4	2.6	9.1	9.6	1.0	8.6	17.8	15.5	13.5	2.0	—
1983g	16.2	13.5	2.7	8.5	11.2	6.5	4.7	16.5	10.5	10.0	0.5	—
				7.6								

a. Coefficients are calculated on the basis of the following information: cols. 1 and 4 = figures of the country in national currency; cols. 3 and 5 = dollars; col. 2 = col. 1 − col. 3. For the foreign savings, the figures used are the deficit in the current account and product in dollars of 1970, converted to current dollars by means of the implicit deflator of the Gross National Product of the United States.
b. GDS = Gross Domestic Savings.
c. GNP = Gross National Product.
d. GNS = Gross National Savings = GDS − FS.
e. FS = Foreign savings.
f. GNP_{fin} = Gross National Product generated by financial institutions, insurance, real estate, and indirect services to firms.
g. Preliminary figures.
Sources: ECLAC, on the basis of official figures. Chile: Central Bank of Chile, *Indicadores económicos y sociales*, 1960–1982.

Table 3. Southern Cone: Monetary system indicators and credit to the private sector

	Argentina						Chile						Uruguay					
	Credit to the private sector		Credit	M₁ᵃ	Quasi-moneyᵇ	M₂ᶜ	Credit to the private sector		Credit	M₁	Quasi-money	M₂	Credit to the private sector		Credit	M₁	Quasi-money	M₂
	Nominalᵈ (1)	Realᵉ (2)	GNPᶠ (3)	GNP (4)	GNP (5)	GNP (6)	Nominalᵍ (1)	Real (2)	GNP (3)	GNP (4)	GNP (5)	GNP (6)	Nominal (1)	Real (2)	GNP (3)	GNP (4)	GNP (5)	GNP (6)
1970	2.0	24.1	22.7	17.0	13.6	30.7	0.01	1.94	7.1	10.2	7.1	17.3	83	1,150	13.6	14.6	6.9	21.2
1971	2.9	25.9	21.8	15.0	11.3	26.3	0.01	2.80	7.9	12.8	4.2	17.0	123	1,353	16.7	18.3	8.4	26.8
1972	4.5	25.3	20.5	15.0	9.5	24.5	0.02	2.86	8.5	13.7	3.8	17.5	253	1,602	20.4	16.0	9.5	25.5
1973	7.4	25.8	20.3	18.4	11.2	29.6	0.09	2.37	7.0	10.6	2.3	12.9	414	1,331	16.3	14.1	7.2	21.3
1974	11.8	33.3	24.5	23.7	13.1	36.7	0.58	2.71	5.9	5.3	1.1	6.4	841	1,525	18.3	12.7	7.3	20.0
1975	29.9	29.9	20.5	23.4	5.4	28.9	3.07	3.07	8.3	4.5	2.6	7.1	1,596	1,596	19.1	11.5	9.6	21.1
1976	121.1	22.3	15.1	16.9	7.2	24.1	13.11	4.20	10.2	3.9	3.7	7.6	2,706	1,797	20.8	12.3	14.4	26.7
1977	433.4	28.9	20.8	16.2	15.3	31.5	49.65	8.29	17.3	4.5	6.0	10.5	4,919	2,064	24.7	11.1	20.0	31.1
1978	1,218.0	29.5	23.7	15.3	18.6	33.9	114.43	13.65	23.5	4.8	7.7	12.5	8,678	2,519	28.4	13.0	25.3	38.3
1979	4,002.3	37.3	28.8	13.3	22.6	36.0	200.90	17.96	26.0	4.9	9.7	14.6	19,109	3,325	34.7	12.3	27.0	39.3
1980	8,344.9	38.7	29.6	12.8	21.2	34.0	379.25	25.09	35.2	5.5	10.7	16.2	34,332	3,654	36.5	10.6	29.1	39.8
1981	22,197.0	50.4	40.7	11.3	22.8	34.1	546.29	30.20	44.5	5.0	14.6	19.6	48,267	3,833	38.2	8.4	35.2	43.6
1982	68,928.0	59.2	43.2	10.2	18.6	28.8	879.30	44.21	71.6	6.6	25.3	31.9	95,255	6,355	74.2	9.2	64.3	73.5
1983ʰ	290,522.0	56.2	33.2	8.6	17.7	26.3	956.06	37.79	63.4	6.8	20.5	27.3	111,380	4,980	59.6	7.0	40.5	47.5

a. M_1 = Bills and coins in circulation plus demand deposits.
b. Quasi-money = time and savings deposits.
c. M_2 = Quasi-money + M_1.
d. Nominal credit, Argentina and Uruguay = Millions of current pesos.
e. Real credit = pesos of 1975.
f. GNP = Gross National Product.
g. Nominal credit, Chile = Billions of current pesos.
h. Preliminary figures.

Source: ECLAC, on the basis of the International Monetary Fund, *International Financial Statistics*, 1982 Yearbook and vol. 37, no. 8, 1984. Credit to the private sector of the monetary system, line 32D; money, line 34; quasi-money, line 35, Central Bank of Chile, *Indicadores económicos y sociales*, 1960–1982.

Table 4. Global external debt of Argentina, Chile, and Uruguay (in billions of dollars)

	Argentina						Chile						Uruguay					
	Private	Public	Total	NET[a]	D/GNP[b]	D/EXP[c]	Private	Public	Total	NET	D/GNP	D/EXP	Private	Public	Total	NET	D/GNP	D/EXP
1970	1.8	2.1	3.9	3.9	13.1	1.85	0.6	2.2	2.8	2.4	30.9	2.24	0.19	0.33	0.52	0.43	16.9	1.79
1971	2.0	2.5	4.5	4.1	13.9	2.14	0.5	2.3	2.8	2.6	27.0	2.48	0.22	0.39	0.61	0.52	19.4	2.44
1972	2.7	3.1	5.8	5.2	16.8	2.52	0.4	2.6	3.0	2.9	28.2	3.05	0.23	0.54	0.77	0.68	23.9	2.19
1973	2.8	3.4	6.2	4.8	16.5	1.68	0.7	2.9	3.6	3.5	33.8	2.46	0.18	0.54	0.72	0.64	20.5	1.76
1974	3.4	4.6	8.0	6.6	10.4	1.74	0.8	3.6	4.4	4.3	37.7	1.89	0.22	0.74	0.96	0.98	24.3	1.92
1975	3.9	4.0	7.9	7.3	16.8	2.26	1.1	3.6	4.7	4.8	42.3	2.56	0.17	0.86	1.03	1.10	22.8	1.87
1976	3.1	5.2	8.3	6.5	16.8	1.80	1.0	3.5	4.5	4.4	37.2	1.86	0.17	0.96	1.13	1.02	22.8	1.62
1977	3.6	6.0	9.7	5.6	17.4	1.47	1.3	3.9	5.2	4.8	36.9	2.00	0.29	1.03	1.32	0.98	24.7	1.63
1978	4.1	8.4	12.5	6.5	21.6	1.67	2.0	4.7	6.7	5.4	40.9	2.28	0.33	0.91	1.24	0.63	20.1	1.36
1979	9.1	10.0	19.0	8.5	28.3	2.07	2.4	5.1	8.5	5.9	44.1	1.84	0.67	1.01	1.68	0.99	22.9	1.41
1980	12.7	14.5	27.2	19.5	36.5	2.75	6.0	5.1	11.1	6.5	49.0	1.86	0.97	1.16	2.13	1.32	35.6	1.40
1981	15.6	20.0	35.7	31.8	46.6	3.31	10.1	5.5	15.6	11.0	59.7	3.10	1.66	1.47	3.13	2.29	34.7	1.83
1982	15.0	28.6	43.6	40.5	58.0	4.84	12.0	5.2	17.2	14.5	72.0	3.70	1.55	2.71	4.26	3.92	49.0	2.77
1983[d]	—	—	45.5	42.1	56.0	4.85	10.5	7.0	17.5	15.5	71.0	3.79	1.31	3.20	4.51	4.11	53.0	3.26

a. Global external debt minus net international reserves.
b. D = Total global debt. Figures for GNP, available in 1970 dollars, were converted to current dollars using the implicit deflator of the Gross National Product of the United States. The figures given are percentages.
c. EXP = Exports of goods and services.
d. Preliminary figures.

Sources: Argentina: Central Bank, *Memoria anual.* Chile: 1970 to 1976, Central Bank, *Deuda externa 1981* (Santiago, 1982); 1977 to 1981, ODEPLAN, *Informe económico 1982* (Santiago, 1983); 1982 and 1983, Informe del Ministro de Hacienda, 2 July 1984. Uruguay: Central Bank, *Indicadores de la actividad económica y financiera.*

Table 5. Southern Cone: Some indicators of capital inflows[a]

	Argentina[b]				Chile				Uruguay			
	K (1)	K/X (2)	LTK/X (3)	CTK/X (4)	K (1)	K/X (2)	LTK/X (3)	CTK/X (4)	K (1)	K/X (2)	LTK/X (3)	CTK/X (4)
1973	147	4.0	−0.2	0.6	387	26.5	−0.1	12.8	9	2.2	−4.9	5.4
1974	−42	−0.9	−1.3	−1.4	211	9.1	0.9	−4.0	96	19.2	1.2	21.0
1975	205	5.9	−1.2	10.7	211	11.5	−4.1	7.7	136	24.7	3.1	5.6
1976	261	5.7	−1.7	−7.7	200	8.3	1.2	2.8	156	22.4	5.7	11.8
1977	556	8.4	8.1	1.7	737	28.3	7.7	21.4	351	43.4	1.7	25.1
1978	302	4.0	28.1	−16.6	1,857	63.1	39.6	15.3	262	28.7	−0.8	−5.9
1979	4,760	51.9	29.1	14.6	2,261	48.9	30.5	−10.2	453	37.9	1.8	7.9
1980	2,176	22.0	31.1	−20.4	3,344	56.0	38.0	16.8	811	53.1	0.9	20.4
1981	1,520	14.0	63.9	−76.0	5,008	90.9	69.7	20.1	494	29.5	2.8	19.1
1982	1,809	20.1	5.8	−18.9	1,096	21.8	24.5	−11.8	−182	−25.9	4.5	33.3
1983[c]	1,570	16.7	—	—	693	15.1	1.3	−5.4	111	8.0	—	—

a. Column 1: millions of dollars; columns 2, 3, and 4: percentages.
b. K = Balance in the capital account of the balance of payments; LTK = Long-term private capital; CTK = Short-term private capital; X = Exports of goods and services.
c. Preliminary figures.
Source: ECLAC, on the basis of information from the International Monetary Fund, *Balance of Payments*.

declined in the neo-conservative period in Chile, increased but marginally in Argentina, and increased significantly only in Uruguay (table 1). In fact, the three countries showed important signs of substitution of foreign for domestic savings in the years 1979–1981 (table 2), which helps explain why investment did not increase markedly during the neo-conservative period. The most striking case of such substitution is that of 1981 Chile. In that year external savings rose from 6 to 13 per cent of GNP whereas domestic savings fell from 11 to 1 per cent.

Fifth, as expected, upon their liberalization, interest rates rose, from being systematically negative during the years of financial repression to generally positive (see table 6). Indeed, borrowing rates proved to be unexpectedly and dangerously high for most of the neo-conservative period: these averaged, in real terms, 41 per cent per year in Chile (1975–1981), 17 per cent per year in Argentina (1977–1980), and 15 per cent per year in Uruguay (1977–1982) for the period ranging from financial liberalization up to the maxi-devaluations.[10] While no one can specify exactly what the equilibrium interest rate is, to judge from other countries' experiences or even LIBOR (which never exceeded 6 per cent in real terms during the neo-conservative period, and averaged much less), it is hard to believe that this could be much above, say, 10 per cent per year real.

The spread between domestic borrowing and deposit rates averaged at least 13 per cent per year in the three countries for the period in question (see table 7). The cost of reserve requirements explains a small part of this differential,

Table 6. Interest rates

	Lending rate			Deposit rate		Domestic debtor		Foreign creditor
	Nominal	Real[a]	Real[b]	Nominal	Real[a]	$l_r(1)$[c]	$l_r(2)$[d]	$l_r(3)$[e]
Argentina								
1971	17.73	−15.4	−20.55	12.98	−18.78	−4.2	−10.1	−9.6
1972	25.58	−23.5	−28.64	19.60	−27.14	−35.8	−40.1	19.6
1973	22.54	−14.8	−6.29	19.27	−17.04	−24.0	−16.5	19.3
1974	22.67	−12.4	−9.89	16.73	−16.66	−20.8	−18.5	16.7
1975	40.89	−67.6	−68.57	20.28	−72.36	199.5	190.7	−90.1
1976	70.02	−62.0	−65.04	56.02	−65.14	6.3	−2.1	−65.4
1977[f]	79.18	−23.3	−22.12	60.50	−31.30	135.9	139.6	−69.1
1977[g]	236.35	15.9	26.70	171.89	−6.34	−14.5	−6.5	16.1
1978	172.35	0.9	11.92	130.41	−14.61	−32.3	−25.0	37.2
1979	134.58	−2.2	2.50	117.14	−9.43	−24.7	−21.1	34.6
1980	98.26	3.7	25.91	79.41	−4.38	−25.0	−10.6	45.7
1981	175.90	19.3	−1.50	152.80	9.30	83.2	51.2	−30.5
1982	213.50	11.4	−13.50	148.75	−19.70	145.4	84.8	−62.9
1983	—	—	—	272.56	−30.19	−1.4	2.9	−22.2

Table 6 (*continued*)

	Lending rate			Deposit rate		Domestic debtor		Foreign creditor
	Nominal	Real[a]	Real[b]	Nominal	Real[a]	$l_r(1)$[c]	$l_r(2)$[d]	$l_r(3)$[e]
Chile								
1975[f]	331.7	−40.8	−45.7	303.5	−44.9	−31.1	−35.0	132.2
1975[g]	490.3	127.1	84.0	234.5	25.2	11.4	−0.8	96.8
1976	250.7	17.7	39.4	197.9	0.0	−27.4	−14.0	45.4
1977	156.3	39.1	55.3	93.7	5.2	−7.6	3.1	20.7
1978	85.3	35.1	33.4	62.8	18.7	−4.0	−5.0	34.1
1979	62.0	16.6	2.3	45.0	4.4	−7.4	−18.7	26.2
1980	46.9	12.0	14.7	37.4	4.7	−12.8	−10.7	37.4
1981	51.9	38.7	38.1	40.8	28.6	6.4	21.2	40.8
1982	63.1	35.1	16.8	47.8	22.5	77.0	53.1	−21.5
1983[h]	42.7	15.9	14.0	27.9	3.9	6.3	4.5	7.3
Uruguay								
1977	65.7	5.3	14.4	38.3	−12.1	−8.8	−1.0	2.2
1978	73.9	19.1	9.0	47.2	0.8	−2.9	−11.2	12.9
1979	65.5	−9.6	−6.5	43.4	−21.7	−26.6	−24.1	19.5
1980	66.6	16.7	29.5	50.1	5.1	−5.2	5.3	26.8
1981	60.4	23.9	39.6	46.1	12.8	4.2	17.3	26.3
1982	61.5	34.0	21.0	53.3	27.2	174.1	147.4	−30.0
1983[h]	94.4	28.3	11.9	70.1	12.3	−7.2	−19.1	32.8

a. Deflated by Consumer Price Index.
b. Deflated by Wholesale Price Index.
c. $l_r(1) = \dfrac{(1 + \text{Libor})(1 + \text{Nominal devaluation})}{(1 + \text{Consumer price variations})} - 1.$
d. $l_r(2) = \dfrac{(1 + \text{Libor})(1 + \text{Nominal devaluation})}{(1 + \text{Wholesale price variations})} - 1.$
e. $l_r(3) = \dfrac{(1 + \text{Nominal deposit rate})}{(1 + \text{Nominal devaluation})} - 1.$
f. Before the liberalization of interest rates (1st semester).
g. After the liberalization of interest rates (2nd semester).
h. Preliminary figures.
Sources: Central Banks of Argentina, Chile, and Uruguay; International Monetary Fund, *International Financial Statistics*.

especially in the early years of high inflation, in which it was important to control monetary growth and during which the Central Bank paid no interest on such reserves.[11] However, the remainder of this unusually high spread, well above the historic one, which ranged between 3 and 5 per cent, seems to constitute a "quasi-rent." Such a "quasi-rent" would be a sign of insufficient competition in this activity.[12] The almost systematic reduction of this spread in the course of time, even during 1981–1982 when the risk of a major financial crisis was quite high, certainly suggests that it was the increased number of financial intermediaries and the ensuing competition that brought it down.[13]

Moreover, domestic interest rates failed to converge to international ones, as had been expected and hoped for, not even during the period in which the exchange rate was being devalued in programmed and pre-announced fashion, and this notwithstanding the heavy inflow of capital. On the contrary, domestic interest rates, both deposit and borrowing rates, proved to be well above LIBOR plus the rate of devaluation. This spread varied from a minimum of 10–20 percentage points per year in Uruguay to 20–30 in Chile and a maximum of 30–40 percentage points in Argentina (see table 7, spreads 2 and 3). Except for exchange risk, this differential made it quite attractive for foreigners to bring capital into these countries and for nationals to borrow in foreign currency (table 6, especially the three final columns). These spreads were large, even in the period when the exchange rate was being devalued far more slowly than the difference between domestic and international inflation, and for this reason one might have expected that real domestic interest rates would be *lower* than international ones, or even negative, at least so long as there was no fear of a maxi-devaluation.[14]

Sixth, financial liberalization was accompanied by a more permissive regulatory environment, both in terms of practices formally permitted and, more importantly, the condoning of practices which though circumventing the spirit of the regulation adhered to its letter (e.g. triangular lending operations, often through firms created for that purpose, whose principal equity was the shares of their parent firms but who, because they were a "different" firm, could borrow on that basis and then relend to the parent firms, thereby circumventing bank regulations on lending limits to any single firm; or restrictions on the ownership of banks by any single stockholder, which nevertheless allowed effective control of a bank to fall into the hands of a same economic conglomerate through the ownership of other persons or firms related to or belonging to the same group). Since, in practice, both the public and the managers of financial intermediaries thought that there was an implicit government guarantee against bankruptcy, there was a built-in bias for banks to engage in excessively risky, but possibly high-pay-off, lending operations, especially to firms belonging to the same conglomerate as the bank, resulting in banks having unduly leveraged ratios of debt to equity. This would be fine in upswings (when the gains were "privatized") but would leave banks "holding the bag" in downswings (or the government would pay the cost if it guaranteed deposits, thus socializing the debt).

Seventh, asset prices (our data are in this case limited to Argentina and Chile) moved quite erratically (see figure 1). The index of stock market prices (expressed in real terms) in Argentina varied by a factor of 1 to 4 (and then back

Table 7. Interest rate differentials (spreads)

	Argentina			Chile				Uruguay		
	1[a]	2[b]	3[c]	1	1A[d]	2	3	1	2	3
1971	4.2	−15.2	−11.6							
1972	5.0	13.4	19.1							
1973	2.7	9.2	12.2							
1974	5.1	5.2	10.5							
1975	17.1	−90.8	−89.2	7.0[e]		117.0	132.2			
1975				78.9[f]		83.9	228.9			
1976	9.0	−67.2	−64.3	17.8	44	37.7	62.1			
1977	11.6[e]	−70.9	−67.5	32.3	16	13.9	50.6	19.8	−3.6	15.5
1977	23.7[f]	9.5	35.5		28					
1978	18.2	26.2	49.2	13.8	13	23.3	40.4	18.1	3.8	22.7
1979	8.0	20.2	29.9	11.7	9	12.7	25.9	15.4	6.7	23.2
1980	10.5	27.4	40.8	6.9	6	20.1	28.5	11.0	10.9	23.0
1981	9.1	−40.3	−34.9	7.9		20.8	30.4	9.8	8.4	19.0
1982	26.0	−67.3	−58.8	10.7		−30.8	−23.7	5.1	−56.3	−54.0
1983[g]	—	−85.2		11.6		−18.0	9.0	14.3	−5.6	38.2

a. Spread 1 = Differences between loan rate and deposit rate.
b. Spread 2 = Differences between interest in US$ received by the foreign creditor who lends in the internal market and LIBOR rate, that is:

$$\left[\frac{(1 + \text{Deposit rate})/(1 + \text{Variation in R})}{(1 + \text{LIBOR})}\right] - 1.$$

c. Spread 3 = Difference between the interest paid by the domestic debtor who borrows in the internal market and the interest paid if he borrows in foreign markets, that is:

$$\left[\frac{(1 + \text{Lending rate})}{(1 + \text{LIBOR})(1 + \text{Variation in R})}\right] - 1.$$

d. Spread 1A = Includes the cost of legal reserve requirements.
e. January–June.
f. July–December.
g. Preliminary figures.
Source: ECLAC. For spread 1A, R. Ffrench-Davis and J.P. Arellano, "Apertura financiera externa: La experiencia chilena en 1973–1980," *Estudios CIEPLAN No. 5*, 1981.

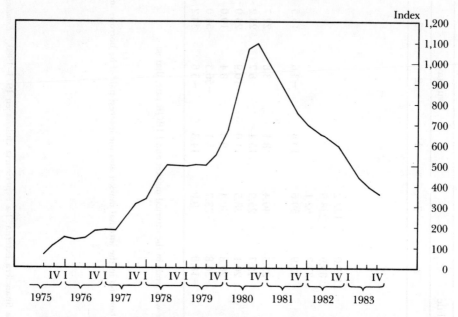

Fig. 1a. Chile: Index of real stock prices, 1975–1983 (December 1969 = 100)

again) in the period between financial liberalization and the maxi-devaluations (mid-1977 to the end of 1980); and the same index rose by more than 1 to 10 in Chile, only to lose half of its value in the same reference period (mid-1975 to mid-1982). Urban real estate seems to have shown similar, though far less extreme, swings. Thus enormous gains and losses in wealth were made during this period. Somewhat paradoxically, to the extent that the prices of shares and of urban real estate are good indicators of asset values, financial liberalization (higher real interest rates) was accompanied by increases, not decreases, in asset values; the subsequent sharp declines were associated not so much with higher interest rates as with the growing gap between "paper" wealth and the ever dimmer prospects of income growth – a prelude, to be sure, of the sharp recessions these countries were about to suffer and the domestic financial crisis which preceded these recessions.

Three Policy Issues

Many questions emerge on viewing the results of economic liberalization. I should like to address three, the answers to which, I believe, can shed much light on the financial crisis that finally ensued:

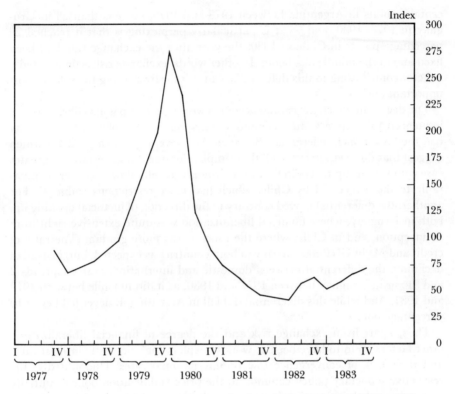

Fig. 1b. Argentina: Index of real stock prices, 1977–1983 (July 1974 = 100)

1. Why were domestic interest rates so high, for so long, and why were they well above international rates despite such high capital inflows?
2. Why didn't domestic savings and investment rise significantly (except for Uruguay) despite the unusually high interest rates, and despite the very strong increase in time and savings deposits?
3. Why did asset prices rise so much, rather than falling when interest rates rose, as we would normally expect?

The Causes of High Interest Rates

The perception of a high exchange risk explains part of the differential between domestic and international interest rates in 1980 in Argentina, in 1981 in Chile, and in 1982 in Uruguay. Nevertheless, the differential had been high in the preceding years when the exchange risk was virtually non-existent and when capital flows were quite heavy. The differential between LIBOR adjusted by devaluation and domestic lending rates (spread 3) was never less than 30 percentage points in Argentina between 1978 and 1979, nor less than 23 in Uruguay in 1979–1980; and what is particularly perplexing is that it reached 29 percentage points in Chile in 1980, the year after the exchange rate had been fixed and in the middle of a boom. In other words, exchange risk is undoubtedly a factor contributing to this differential, but it is far from being the sole or most important one.

The degree in which the capital accounts were opened up is also likely to have influenced this interest-rate differential. This may well explain why Uruguay had the lowest such differential. But even then several problems remain which suggest that the relation is not all that simple. For one thing, despite the greater ease with which capital could flow into Uruguay, capital flows to it were comparable to those received by Chile, which had more restrictions (table 4). For another, the differential moved contrary to the direction of financial opening up, both in Uruguay, where financial liberalization was quite extensive right from its inception, and in Chile, where the process was more gradual. The ratio of credit and M_2 in GNP rose sharply in both countries as expected, but, instead of declining, the differential between domestic and international rates (spreads 2 and 3) *rose* in Uruguay between 1977 and 1980, as it did in Chile between 1979 and 1981. And while this differential did fall in Argentina, it never fell below 10 percentage points.

Thus, apart from exchange risk and the degree of financial liberalization, other factors seem to have been at work to explain high domestic interest rates and their lack of convergence toward international rates. Undoubtedly, the restrictive monetary policy common to the price stabilization policies initially pursued in all three countries was one such factor, which would help explain high interest rates in the early years of these experiences. Yet, since these interest rates continued to remain high in real terms, even in periods of very heavy capital inflows and when stabilization policy moved from controlling the money supply to controlling the exchange rate, it is reasonable to look for additional explanations on the side of the demand for credit. Among the principal factors which increased this demand above normal levels at different times during the neo-conservative experience were:

1. The unexpected appearance of opportunities for exceptional capital gains, something which will naturally raise the real demand for credit. This is what happened in Argentina in 1977, when as part of one more attempt at price stabilization, a four-month price freeze was announced. This encouraged firms to demand credit to buy inputs and stockpile output in order to sell it later at the higher prices expected once the price freeze was lifted. The most notable case is that of Chile in 1975, when in the midst of a severe depression the government announced its programme to auction off a large number of banks and enterprises which had come into its hands during the Allende government.

2. The rise in the real value of assets. Under normal conditions the rise in such values is closely related to the rate of economic growth which is soon

expected. Nevertheless, during various years after financial liberalization the sharp upward revaluation of stocks and real estate far exceeded what could be justified by any reasonable expectation of likely economic growth, indicative of a speculative euphoria (a point to which I shall soon return). In any case, whatever the reason the real market value of assets did grow substantially for several years, at least in Argentina and Chile. To the extent that this happens, the demand for credit can rise, because of a wealth effect.

3. Changes in the ways in which public enterprise deficits were financed. In the past, these had been largely "financed" via the direct creation of money by the Central Bank; from now on, they were financed by borrowing in the domestic capital market. Chile used this mechanism quite extensively, especially at the beginning, whereas Argentina and Uruguay borrowed heavily from external sources to cover public sector needs throughout the entire process.

4. The opening up of trade and the elimination of administrative controls on the allocation of credit. As a result of these measures, the demand for consumer credit expanded immensely, especially that for consumer durables, for the relative price of these had fallen considerably within these countries because of lower tariffs.

5. The belief that high real rates of interest were transitory and that they would soon fall to reasonable (equilibrium) levels, say 5–10 per cent per year. This proved to be all the more important because much short-term credit was being utilized to finance operations of a longer-term nature, thereby supposing the automatic renewal of credit.

6. The above are all factors which explain a demand for credit originated in expectations (correct or not) of higher future income. One would suppose that the demand for credit would fall should expectations be reversed. Nevertheless, it is important to point out that the demand for credit can also go up in the short run, not to take advantage of possible gains in income but to *avoid* or postpone possible losses in wealth brought on by unexpected reversals in key economic indicators: for example, to avoid (a) the hurried sale of excess inventories accumulated because of the unexpected decline in sales, or (b) the forced sale of assets during periods of recession and consequently at depressed values. It will thus be quite tempting to postpone such capital losses if the recession is considered to be transitory and it is thought that sales and/or assets will soon recover their expected values (Argentina between the end of 1977 and the end of 1978, and Chile in 1974–1975). The temptation will naturally be irresistible should the sale of assets – because it is a generalized situation affecting many firms, as was the case after the maxi-devaluations of 1981–1982 – have to be made at such low prices that it implies the firm's bankruptcy. For the path to bankruptcy need be neither smooth nor gradual, nor indeed is it always evident just when it need take place. And when such a process is sudden and widespread (as in 1981–1982), it is apt to induce the acquiescence of bank creditors, for their own solvency is at stake (a point to which I shall return).[15]

This introduces an important *assymetry and upward bias* in the demand for credit. For to the extent that potential capital losers demand more credit, and not less, in order to avoid or put off losses, the increased demand for credit of would-be capital gainers is not compensated by a decrease in the demand for credit of prospective capital losers. Indeed, rather than cancelling each other out, these effects combine and reinforce each other, and the overall effect is all the stronger, the greater the fluctuations in the perceived value of capital assets. Precisely one of the central features which characterized the neo-conservative experiences was the sharp changes in the relative price structure: of prices with respect to wages; of agricultural prices with respect to manufacturing; of the price of tradeables with respect to that of non-tradeables – all of which necessarily gave rise to important capital gains and losses.[16] To the extent that capital losses in particular were perceived as transitory – excusable, for to perceive them as permanent might well imply recognizing insolvency, all the more so since key macro-economic variables fluctuated substantially and never approximated equilibrium – an important asymmetry was introduced in the demand for credit, wealth transfers leading both potential capital gainers and losers to demand more credit, consequently moving real interest rates well above equilibrium levels.

In much the same vein, the failure of domestic interest rates to converge to international ones was due to the fact that the demand for credit grew far more than the international market was willing to finance. This latter market is rationed by quantity as well as price: while certainly an interest-rate differential can attract capital inflows,[17] it will do so in practice as long as the exchange risk is low. So long as the demand for foreign credit increased in step with the capacity to service such debt, that is to say exports, reserves and similar indicators, exchange risk was likely to be perceived as low. However, once the increased demand for credit was due not to factors which were related to the increased capacity to service such debt, but rather, as occurred during the end of phase 2, to a deterioration in the capacity to service such debt (because of the lag in the exchange rate, the international recession, and growing interest payments), the supply of foreign credits was sharply cut back. As might be expected, capital inflows then became rather insensitive to interest-rate differentials but quite sensitive to exchange risk. Domestic interest rates were thus pulled upwards, worsening the recession in each country, and ultimately precipitating an acute financial crisis, all of which forced a maxi-devaluation.

For this reason, domestic and international interest rates failed to converge, let alone equalize. To be sure, had it been possible to maintain this exchange policy indefinitely, such interest-rate convergence would eventually have taken place. But the point is that the longer the pre-announced exchange-rate policy was maintained the more it became overvalued, and the less credible its continuance became. For it was hard to believe that the government would be willing to persevere in its exchange policy however sharp an economic contraction it required. No such guarantee could be given, short of actually closing the Central Bank, counting on an *indefinite* supply of foreign exchange, and thus "dollarizing" the economy, as in Panama.

The Causes of High Financial Savings and Low Real Savings and Investment

Financial liberalization was expected to raise interest rates and encourage greater savings and investment. Interest rates rose, indeed, as did time and savings deposits. Yet except for Uruguay, "ex-post" national savings (i.e. income less consumption) did not increase, nor did investment. Why didn't the increased financial savings translate themselves into increased effective national savings and investment?[18]

Precisely because interest rates were not free but controlled, "financial repression" in the Southern Cone required administrative forms of allocating that credit. It would seem, especially with the benefit of hindsight, that credit was heavily biased in favour of fixed capital and public works infrastructure at the expense of consumer credit, private sector infrastructure (i.e. commercial construction), and housing. Whether such an allocation maximizes welfare or not – implying that the social discount rate was less than the free market rate of interest – is an open question. The fact remains that financial liberalization both freed interest rates and eliminated controls on credit use. Thus the "repressed" demand for credit – especially for consumer durables but also for private commercial infrastructure – manifested itself upon "financial liberalization." Hence, the increase in financial savings did not necessarily yield an increase in effective, ex-post savings and investment but rather helped finance consumer credit. This effect was all the stronger in a country such as Chile where – by policy, and unlike Argentina and Uruguay – the government purposely reduced public sector infrastructure investment, so as not to crowd out private investment. In retrospect, it now seems clear that this reduction in public investment gave rise only to a partially offsetting increase in private investment (be it in infrastructure or machinery). For these reasons, overall savings and investment in Chile were actually much lower during the neo-conservative period, and in Argentina were virtually similar to pre-financial liberalization days, notwithstanding the sharp increase in financial savings. Effective (ex-post) national savings thus proved rather insensitive to the rate of interest in this period, at least in Argentina and Chile.

Rather, savings seem to have responded: (a) positively to growth in national income and especially to the ups and downs in national *disposable* income deriving from fluctuations in the terms of trade; (b) inversely to the increase in the availability of domestic credit for consumer durables as well as to the relatively low cost of dollar-denominated credit (at least up to 1981) for imports; (c) inversely to the decline in the relative price of durable goods, especially imported ones (due to the overvalued exchange rate of phase 2 stabilization and the reduction in tariffs); and (d) inversely to the apparently greater market value of most fixed assets, which led economic agents to believe (mistakenly) that their permanent income was higher, so that they could well afford to spend more on consumption.

Thus, whatever the long-term effects of higher interest rates may be, of and by themselves, on domestic savings, the evidence of the Southern Cone is certainly mixed. In Uruguay, real savings and investment rose. In Argentina and Chile, on the other hand, consumption was induced (possibly because it had been heretofore so repressed). In any case, it should be clear that, important as the impact of financial liberalization on interest rates is, so too is the dismantling of credit controls it entails. To the extent that these were biased in favour of investment, the impact of liberalization could be to increase financial savings yet reduce real savings (and investment).

For much the same reason, the increase in "foreign savings" (debt) associated with increased financial opening up to the outside world need not have resulted in like increases in investment. Some debt indeed was used to increase foreign exchange reserves (at least up to the period preceding the maxi-devaluations). Some substituted domestic savings with foreign savings, as noted earlier, especially in the years 1980–1981, giving rise to massive increases in consumer imports. Much went to satisfy a heretofore "pent-up" demand for the import of military hardware (less valued by civilian governments concerned with fostering productive investment). Finally, much augmented private savings and investment *overseas*, as foreign debt was "socialized," whereas foreign exchange was "privatized." The end result is that notwithstanding unprecedented levels of foreign savings in the neo-conservative period – to wit, foreign savings rose with respect to the pre-neoconservative period approximately the equivalent of 1 per cent of GNP in Uruguay, 2 per cent in Argentina and 3 per cent in Chile in these years[19] – overall investment rose by less than that amount in Argentina (1.5 *v.* 2 per cent), and fell in Chile. Only Uruguay showed a marked increase in investment in the period.

The Behaviour of Asset Prices

I have already commented on the extraordinary volatility of asset prices during the period of financial liberalization; not just that they varied far more than any other variable – for that is not unusual – but that they varied by as much as they did: 4 to 1 in Argentina, and over 10 to 1 in Chile. This is all the more puzzling given the noted tendency of interest rates to rise sharply and remain high during the period of financial liberalization. In short, how does one account for such extraordinary increases in the prices of stocks (in real terms) in a period characterized by unusually high interest rates?

Most economic theories prepare us for the reverse results: that stock prices vary inversely with interest rates. Since at higher interest rates future income streams are discounted at ever greater percentages, the present value of such income streams (or stock prices) falls with increased interest rates. This is, of course, true of the more traditional hypotheses regarding the relationship between interest rates, money supply, and stock prices, namely that increased money supply in time t will lead to higher stock prices in t + 1: first, because money supply increases lower interest rates and so raises the present value of future earnings; secondly, because increased money supply may increase aggregate demand, and so real earnings, when there is idle capacity; and, thirdly, because in the short term excess money holdings may be transferred more rapidly into stocks (raising their demand) than into goods or bonds. Any one, and all three together, point to a positive relationship between money supply increases in previous time periods and stock price increases in the same and following time periods, and an inverse relationship between interest-rate behaviour and stock prices in the same or following time periods.[20]

This traditional approach has been successfully challenged by the most modern formulation of stock-market behaviour, the efficient capital market hypothesis,[21] which argues that the price of a stock already incorporates all past and current information concerning the best estimate of future values of its determinants (including the interest rate). Hence, it cannot vary systematically *today* as a response to *past* variations in money supply or interest rates. Nevertheless, even the efficient capital market hypothesis would suggest that, since financial liberalization can be expected to raise interest rates, stock-market prices should fall as liberalization is announced or to the extent that it is expected. This conclusion could be avoided, if one believed, as did those who argued on behalf of financial liberalization, that freeing interest rates would raise not only real interest rates but also the quantity and quality of investment, leading to far greater growth. In fact, in retrospect such growth did not take place. Yet that may only show that buyers of stock *erred* in expecting so much higher economic growth; but their expectation of higher growth led them to raise the demand (and so the price) for stocks, notwithstanding their expectation that interest rates would also rise. Once their expectations of strong growth were dashed, stock prices tumbled down. This explanation fits the general swing in stock prices, though it can hardly explain the *magnitude* of these savings. For example, lending rates rose from 39 per cent per year (real) in 1976 to 55 per cent per year in 1977 in Chile. Such a rise in interest rates implied discounting the earnings of future years so heavily that earnings as of the third year and beyond would have a present value of less than 9 per cent! Obviously, then, for stock prices to rise in 1977 enormous growth would have to have been expected for the years 1977, 1978, and 1979. Chile's growth did accelerate quite strongly, from just under 2 per cent to about 7 per cent per capita in those three years. But when one discounts such growth by 55 per cent in the first year, 80 per cent in the second, and 90 per cent in the third, it is obvious that no reasonable expectation of accelerated growth could compensate the increased interest rates. And yet

Chilean stock prices rose 76 per cent in real terms between the fourth quarter of 1976 and the fourth quarter of 1977.

Thus, such wide swings in stock prices must have reflected both very favourable expectations as to stronger economic growth and the belief that interest rates were transitorily high but would soon fall and settle down to much lower rates.[22] Only in some such way could one rationalize the quadrupling in *real* stock prices between the first quarter of 1978 and the first quarter of 1980 in Argentina or the sixteenfold increase in Chile between the third quarter of 1975 and the fourth quarter of 1980. Optimistic assessments as to future increased economic growth and declines in interest rates leading to such enormous increases in real stock values can only be characterized as generalized euphoria; in short, the upswing of a speculative bubble.

Such a bubble and crash has been ably demonstrated by Meller and Solimano for Chile.[23] I will use the same formal definition: namely, a bubble can be defined as a situation in which the price of shares between two periods (adjusted by dividends) grows faster than the interest rate, and continues to so exceed it, for several succeeding periods, only to be followed by repeated periods in which the growth in the price of shares is slower than that of the interest rate. Such behaviour could be characterized as a speculative bubble followed by a crash, inasmuch as in an efficient market the growth in the price of shares (adjusted by dividends) should equal the interest rates. Thus any growth persistently and *systematically* beyond that explicable by real economic forces (i.e. the interest rate) would be symptomatic of the formation of a speculative bubble – an interpretation which would be corroborated if such growth were followed by a crash (in which stock prices grew far less than the interest rate).

Using this concept, a bubble clearly developed in Chile between the third quarter of 1979 and the end of 1980 (see figure 1a), when the return to stock purchases far exceeded the interest rate for five successive quarters, stock prices (the boom) more than doubling in real terms in that brief spell, after which they declined for the next 12 successive quarters (the crash) to one-third their peak value. Similarly, in Argentina (see figure 1b) stock prices grew in real terms well above interest rates for eight successive quarters from the beginning of 1978 through the beginning of 1980 (during which stock prices grew four times in real terms), to be followed by a decline in the next nine successive quarters (when they fell to one-sixth of their peak value), before they began to recover.

Another way to present the above phenomenon is to note that in neither of the two countries during the period of financial liberalization was the level of stock prices (in real terms) correlated with the interest rate. Indeed, what correlation there was (especially the case of Chile) seems to have been with M_1 (in real terms) and M_2 (in real terms); positively with the former, as if the excess supply of money were spent far more on stocks (thus raising their prices) than on goods, thus raising stock prices in real terms, and negatively with M_2 (real), as if short-term time deposits were good substitutes for stocks.[24] But in neither country was the level of interest rates correlated (negatively) with the level of stock prices, as might have been expected. Nor was there any significant correlation, even in the short run, between quarterly percentage changes in stock prices and variations in the interest rates; and the sign is positive (contrary to the inverse relation that might have been expected).

There are, however, two interesting results (for Chile) related to interest-rate variations. Interest-rate increases in period $t-1$ are followed by a rise in stock prices in period t; moreover, stock prices in period t rise, when interest rates fall *in period $t+1$*.[25] This suggests that interest rates did not affect the level of stock prices in the long run. But in the short run, economic agents behaved as if they thought they did: raising stock prices in time t because they thought interest rates would fall in time t (since they had risen in time $t-1$), and also raising stock prices in time t if interest rates were expected to fall in time $t+1$, for they believed that if interest rates had risen in $t-1$ they would fall in time t of $t+1$.

Thus stock-market behaviour bore no correlation to interest rates in the long run – as it formed a speculative bubble – but in the short run, at least in Chile, variations about this level occurred that were related to what economic agents thought short-run interest behaviour would be. In any case, a bubble did form in stock prices in both Argentina and Chile. Since stock prices reflect the market value of fixed capital, a non-tradeable, all firms and asset-holders generally thought themselves far wealthier in real terms (and especially in terms of tradeables, given the lag in the exchange rate). Since the value of their assets seemed to rise far more than the value of their foreign debt (at the fixed exchange rate) they thought themselves well off, and so capable of paying high domestic interest rates and/or spending more on consumption (because of their presumed wealth gains). Thus the bubble in asset values led to further divergences from equilibrium in the credit and exchange markets.

Consequences: The Bubble Bursts and the Financial Crisis Ensues[26]

For all these reasons, the demand for credit remained strong and domestic interest rates high during the neo-conservative experiences (at least up to the maxi-devaluations). But how can an economy function if its productive sectors are paying real interest rates of the order of 20 per cent per year or more? For it is really quite difficult to imagine that there exists a wide set of investment opportunities that permits the paying of such interest rates for a prolonged period of time if, in fact, average growth rates are as modest as they were throughout most of the neo-conservative period.

Part of the explanation no doubt lies in the fact that not all borrowing took place at domestic rates; much was in dollars at internationl rates. Thus, at least while the lag in the exchange rate persisted, those firms which borrowed abroad paid negative real rates of interest on their loans (table 6), a situation which obviously could continue only as long as the strong inflow of foreign capital continued. Once this slowed sharply, as it did just before the maxi-devaluations, the cost of foreign borrowing proved quite costly. Nevertheless, at least for some time, the inflow of foreign capital at negative real rates of interest for the domestic borrower permitted him to pay high domestic interest rates for those loans contracted within the country.

The bulk of long-term credit was thus in foreign currency, for this was at reasonable interest rates, though variable, and, of course, it always carried an exchange risk. The fact that the bulk of long-term credit was in foreign currency, whereas the bulk of domestic credit was of short-term duration, largely explains why it was virtually impossible for any but a small part of dollar-denominated debt to be transferred into domestic currency debt toward the end of phase 2 when the risk of a major devaluation loomed large.

In practice, access to foreign credit was neither uniform nor generalized. Rather, this proved to be a wholesalers' market, largely limited to big firms or those belonging to the same owners as the banks. Thus, such economic conglomerates would find it much more attractive to buy up assets (from those with less access to foreign or domestic credit) than invest, a fact which tended to raise the value of existing assets all the more. Moreover, the fact that such conglomerates were able to devise mechanisms to invest without significantly drawing on their own limited financial resources, but on that of the community as a whole (through the banking system) gave them great leverage (high debt/capital ratios) which accelerates booms (bubbles) but also accentuates declines (when the bubble bursts).

Indeed, as noted earlier, because of the implicit guarantee against bankruptcy that at least the major banks were thought to have (i.e. that the government would bail them out rather than allow a run on the financial system) and the fact that these banks were often controlled by parent firms desirous of credit, the banking system itself acquiesced in allowing highly leveraged, pyramidal, and triangular lending operations. If all went well (as during upswings) heavy profits would accrue to the most leveraged firms. If not, firms' and banks' losses would be limited to their relatively low equity; the bulk of the loss would of necessity be incurred by depositors – or if, as expected, and as it largely turned out, deposits were guaranteed by governments to avoid a financial run, most of the loss would be absorbed by the public at large.

The generalized increase in the value of assets, a result of exuberant expectations as well as of the massive inflow of foreign capital and the formation of economic conglomerates, implied huge capital gains (at least on paper) for all firms, so that they thought themselves capable of paying such exorbitant interest rates. For, indeed, about the only thing that did grow at rates at all similar to interest rates were asset values, at least for a goodly part of the period from financial liberalization up to the maxi-devaluations of 1981–1982.

Thus, the domestic financial bubble – with its upward revaluation of assets

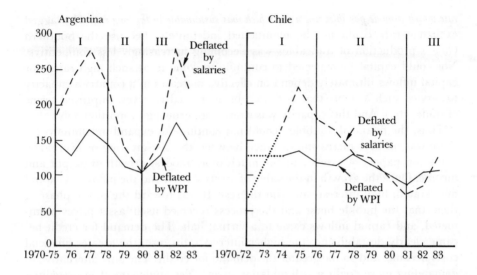

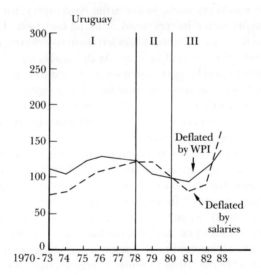

Fig. 2. Real effective exchange rate.

together with high interest rates – could continue to maintain itself only if fed with an ever-increasing supply of foreign credit, attracted, to a large extent, by the highly favourable interest rate differential, itself the result of the lagging exchange-rate policy such flows themselves made possible. *Thus the financial bubble rested on a growing influx of foreign capital and the corresponding lag in the exchange rate which it made possible, neither of which was sustainable in the long run.* The lagged exchange rate could not be maintained indefinitely, because the Southern Cone's production of tradeables was becoming increasingly less competitive. Nor could capital be expected to continue to come in at such high rates, for capital inflows ultimately depend on effective increases in a country's capacity to service such a debt (that is to say, on increased reserves, improved terms of trade, etc.). And this capacity was simply not growing proportionately.

Thus, the financial "bubble" could not continue to expand indefinitely, for there was an ever-increasing distance between the fast-growing or overblown value (on paper) of assets and the much more modest growth of output and income. Once the growth in the value of assets slowed and the inflows of capital decelerated, high interest rates did the rest. It was toward the end of phase 2, then, that the bubble burst and the process reversed itself: asset prices plummeted, and capital inflows came to a virtual halt. The demand for credit became clearly destabilizing: the higher interest rates rose, the greater financial costs became, and so the greater the demand for credit. The only alternative to demanding more credit was liquidating assets. Yet, under the then prevailing conditions, that was tantamount to declaring bankruptcy, for the capital losses would be enormous even if buyers could be found for assets. The obvious reluctance of firms to incur such capital losses led them to demand credit in the hope that somehow something would turn up. At this stage, they really had nothing to lose, and much to gain by postponing such a liquidation of assets. Thus, asset market disequilibrium (an unwillingness to sell at a significant loss) was thrust on the financial market for its resolution, increasing the demand for credit and driving up interest rates even further, the burden of adjustment thus falling on credit markets and the interest rate.

In effect, firms perceived the payment of high interest rates to cover current financial costs as a more attractive alternative than simply taking immediate huge capital losses. Given such short-time horizons it was small wonder that they were willing to "pay" inordinately high interest rates: these averaged 26 per cent (Argentina), 40 per cent (Uruguay), and 58 per cent (Chile) in *real* terms in the year preceding the maxi-devaluations in each. For analogous reasons, the banks (their creditors) tended to go along with firms and renew such credits. For were the banks to refuse to renew credit and try to make good on the guarantees, they knew that these were now worth but a fraction of the overblown values at which they had been assessed when credits were first provided.

In other cases, many of the debtor firms belonged to the same economic conglomerate as the bank itself. In this way, be it for one or the other reason, the banks tended to play along and postpone the forcible liquidation of assets, maintaining credit lines but at ever-higher rates of interest. The consequence of this, naturally enough, was that the solvency of the banks and the financial system as a whole came to be completely jeopardized, for it was overwhelmingly dependent on the financial state of firms whose situation was precarious at best.

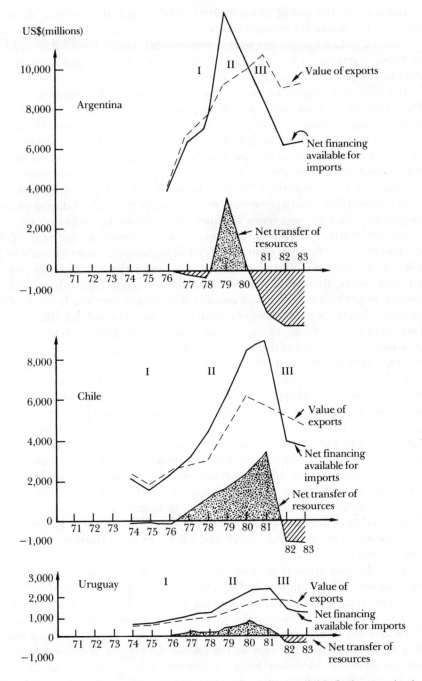

Fig. 3. Exports, net transfer of resources, and financing available for imports (net).

But towards the end of phase 2 (1981–1982), when the domestic financial crisis exploded, and the contractionary effects of the international recession and the overvalued exchange rates were apparent to all, capital flows fell sharply: 39 per cent in Argentina in 1980; 75 per cent in Chile in 1982; and over 100 per cent in Uruguay in 1982.

It is difficult to exaggerate the adverse impact which such a shift in net capital flows implied. Indeed, once interest and other factor payments are deducted from net capital flows, instead of receiving resources from the rest of the world, the three Southern Cone countries became net *exporters* of resources in the year they were finally forced to devalue (see figure 3). The net transfer of resources was *negative* and of the order of 20 per cent of exports in all three countries, after having been strongly positive the year before. Indeed, it was because of such strong capital flows that aggregate demand could be maintained during phase 2 despite the generally poor terms of trade and the lag in the exchange rate.

Put differently, the shift in net resources transferred in the year of the devaluation was the equivalent of a deterioration in the terms of trade of 25 per cent in Argentina, 50 per cent in Uruguay, and 80 per cent in Chile (fig. 3). In other words, this meant, for example, that instead of Chile's being able to import 50 per cent *more* than the amount of its export earnings because of the positive effect of the net transfer of resources – as in 1981 – in 1982, when the net transfer of resources was negative, it had financing available which only allowed it to import 73 per cent of the value of its export earnings. Hence, capital flows were abruptly reduced just when they were most necessary (the end of phase 2 and the beginning of phase 3), and were exaggerated (most of phase 2) when they were far from indispensable. Thus, rather than helping smooth the business cycle, capital flows accentuated it, proving to be highly pro-cyclical.

This financial "run" on the part of foreign banks severely aggravated the contractionary effects of the overvalued exchange rate, of the international recession (on export volumes, and terms of trade and interest rates) and of the domestic financial crisis. Moreover, once capital flows were cut back there was no longer any confidence in the sustainability of the exchange policy, for resources (reserves) had finally been depleted to finance it. Hence, there was no longer any other practical alternative but to abandon the policy of gradual and pre-announced devaluations, which had led to the overvalued exchange rate, and devalue massively.

Given the magnitude of the disequilibria, and the brief time-frame in which external accounts had now to be brought into line, adjustment could only be of the worst type – solely contractive (output reducing) rather than expansive (output switching). Consequently, in the two or three years which followed, GNP fell some 10 per cent in each of the three countries (as opposed to a reduction of the order of 4 per cent in the rest of the region) and unemployment sharply increased. Moreover, given the severity of the domestic financial crisis, the Central Banks of the three countries were finally obliged to step in and intervene directly or indirectly to support the domestic banking system, renegotiate or write off much of the firms' domestic debts, as well as to renegotiate foreign debt, both that with and that without public sector guarantees.

Finally, notwithstanding the fact that the inflow of foreign capital was sharply curtailed in these years, the level of foreign debt was still extraordinarily high by the end of 1983. The ratio of foreign debt to the value of all exports of goods and services varied from a low of 3.2 in Uruguay to a high of 4.9 in Argentina. This, of course, compared quite unfavourably with the average of 2.6 for the rest of the region. To be sure, the Southern Cone countries had also been amongst the most highly indebted countries of Latin America when the neo-conservative experiences began. What is truly remarkable is that they should not have slowed down their indebtedness in the course of eight to ten years of considerably strong export growth and seeming allegiance to the principles of strict financial discipline. That they should still stand out amongst the most indebted countries of the region in 1983 certainly does not speak well of the economic liberalization policies which they pursued, and, in particular, of their policy of financial liberalization. This latter seems to have heightened rather than reduced their dependence on foreign savings and consequently made them all the more vulnerable to swings in the international economy. For now they had to be prepared to offset unexpected movements in capital accounts as well as in their terms of trade. If their debt had been more modest, financial liberalization might have given them the freedom to cope with the external disequilibrium they faced in later years. Instead, rapid financial liberalization, in the face of an already unduly high level of debt and in the presence of major domestic disequilibria, as evidenced by abnormally high interest rates, added a further and critical element which would serve to accentuate rather than attenuate unexpected movements in their external accounts. Consequently, they lost rather than gained a degree of freedom. Adjustment was thus largely forced upon them (maxi-devaluation plus severe recession) rather than being a policy which they deliberately chose. The overindebtedness of phase 2 thus eventually led to the capital "flight" and overadjustment of phase 3.

Conclusions, Theoretical Implications, and Policy Lessons

1. There is no doubt that at the beginning of the neo-conservative experiences, the domestic capital market was quite repressed and underdeveloped. Nevertheless, the profound changes which financial liberalization and opening up brought about did not translate themselves, despite intentions, into systematically higher savings or into clearly improved resource allocation. Indeed, the three experiences came to a close with their financial systems in a shambles.

2. The principal failing seems to have been in the persistently high real rate of interest throughout almost all of the neo-conservative period, a rate of in-

terest which far exceeded the rate of growth of output or any reasonable rate of return on productive assets. Real interest rates of the order of 2–3 per cent per month, as was the case in all three countries during a good part of the period, cannot be paid systematically without jeopardizing the solvency of firms and ultimately of the financial system itself. The bursting of the financial bubble was, thus, inevitable. Unlike "normal, well-behaved" markets where divergences from equilibrium automatically set in motion forces to restore that equilibrium, deviations from equilibrium in financial markets may lead to even greater divergences, if certain minimum thresholds of confidence in the ability to service such debt are not met. Once such confidence is lost, incentives are set in motion which may lead to creditors demanding more credit to postpone insolvency and to banks acquiescing for fear of having to take losses which may exceed their own reserves. In such conditions, divergences from equilibrium (high interest rates) may lead to behaviour which further accentuates such divergences (raises interest rates further). The system comes to rest, then, only once the "bubble" bursts, and a major financial crisis erupts, as was the case in the Southern Cone.

3. The real significance of high interest rates, even in the early stages, was incorrectly interpreted by the authorities. Rather than see it as a sign that something serious was amiss in the workings of the economy – that it was a sign of a possibly major disequilibrium – they tended to rationalize high interest rates away, considering that inasmuch as it was the rate that equalized the supply and demand for credit, it was by definition the equilibrium rate. This, of course, was a major theoretical error: a confusion of the market clearing rate of interest with the equilibrium rate of interest. For the equilibrium rate of interest is that which equalizes supply and demand when *all other markets* (asset, foreign exchange, labour, and good markets) are also in equilibrium. If these other markets are not in equilibrium, the rate of interest which then clears the credit market is not the equilibrium rate of interest; rather it is the rate required to absorb the disequilibrium of other markets. Such was the case during the bulk of the neo-conservative period. The high rate of interest was a reflection of disequilibria in other markets: at different points of time, it was a problem of the market for foreign exchange (due to the overvaluation of the exchange rate and the expectations of devaluation), or of the market for goods (inflated prices) or of assets (the bubble). As a result, a good part of the disequilibria in these markets was thrust on the credit market for its resolution, inasmuch as this is a relatively fast, price-adjusting sector.[27] This analysis thus confirms the views of those who insisted all along that such unusually high real interest rates were indicative of a basic disequilibrium in the economy, and rejects the views of those who argued in somewhat "panglossian" fashion that if the market so dictates, then they are the correct ones. It was thus a grave policy error to have liberalized financial markets so rapidly, and to such an extent, precisely at a time when, because of the stabilization policy, important disequilibria still remained to be resolved in other critical sectors of the economy.

4. Domestic and international interest rates failed to converge, much less equalize, because credit markets have important peculiarities. Credit cannot be efficiently rationed solely by price (rate of interest) because credit is a future commitment. Hence, the higher its price, the less is the belief in the debtor's capacity to meet this commitment. Thus, in practice, credit must be rationed both by quantity and by price, which means that capital inflows will be sensitive not only to interest rate differentials but to the amount demanded. In so far as all of these other markets were transferring their disequilibria upon the credit market for resolution, the amount of credit, and therefore capital inflows, required in order to equalize domestic and international interest rates was enormous, far in excess of what would have been demanded had these other markets been in equilibrium. It is not strange, then, that international banks were not willing to lend the full amount demanded, thus preventing the law of one price from fully operating in the financial market. Indeed, it is now quite clear, with the benefit of hindsight, that more foreign capital came in than was in fact prudent from a long-run perspective.

5. The fact that credit, especially foreign credit, was rationed by quantity and not just by price, gave an additional advantage to those firms that belonged to economic conglomerates, or were themselves large, or were dedicated to exports, for they had much better access to this rationed, but cheaper, foreign credit. Those firms with access to international capital markets (or related to banks with such access) had the privilege of bringing in capital to the country at negative real rates of interest (in terms of domestic currency) for a long period of time and then relending it, in domestic currency, for short periods of time and at high real rates of interest, or using it to buy assets at good prices (to the extent that other asset-holders only had access to credit at high interest rates), obtaining in this way substantial profits.[28]

Such privileged access for some was not due to legal discrimination but was a reflection of reality as such. Capital markets were (and still are) segmented. International capital markets are largely "wholesale markets," with access naturally restricted in practice to the principal firms and banks of a country (or to firms linked to such banks or to the export sector). Thus, most small and medium-sized firms, as well as those in the production of non-tradeables (such as construction) or whose production was geared primarily to the domestic market, found themselves restricted largely to the domestic credit market to satisfy their needs (be it in domestic or foreign currency) but paying high interest and/or intermediation charges.

In short, much as liberalization stimulated financial intermediation, the capital market remained largely segmented and underdeveloped, especially as far as long-term credit was concerned. For this reason, too, it proved very difficult to raise savings and improve resource allocation. It should have been foreseen that, because of the inevitable rationing of credit, access to foreign credit would have been differentially available to the different firms. This would have provided additional justification for introducing some measure of direct intervention into this market to control the rationing of credit and to redress this type of segmentation.

To be sure, to borrow in dollars was to run the risk of an unexpected devaluation, a risk that was to prove all too real in phase 3 when each was forced to realize a maxi-devaluation. Nevertheless, as this risk was seen as rather remote in the beginning of phase 2 (and it was), the incentive to borrow abroad was enormous, almost irresistible. Therefore, by the time the accumulated overvaluation had become unsustainable and the exchange risk was high, the accumulated stock of foreign debt had become quite large; hence the impact of the much-needed maxi-devaluations on debtors in foreign currency was devastating, many of the gains of previous years having been wiped out.

6. The new capital market was almost exclusively limited to short-run instruments. It would have been wiser in retrospect to have inverted the order, first generating long-run instruments (indexed) for various years' duration and with good interest rates, and paying low interest rates to depositors for short-run money. For it could have been foreseen that were the market to be left to itself – given the need for credit and the environment of uncertainty and strong inflation – it would naturally tend to create short-run instruments, and at high interest rates. Once such short-run instruments had established themselves, it would be very difficult for long-run ones, especially bonds, to emerge, for these require stability and predictability; in other words, they require other markets to be at or close to equilibrium, which of course they were not.

7. There were important differences between the three countries as far as the legal controls and limits on the entry and exit of capital were concerned, as well as in the timing and sequence of financial liberalization relative to trade opening up. Yet such differences do not seem to have been of decisive importance in explaining differences in the rate of capital inflows (strongest in Chile despite its greater controls), nor in explaining the failure of interest-rate convergence to take place (it failed to materialize by and large in all three countries). More important were the factors that I have pointed out, related to the stabilization policy and to the demand for credit.

Also overlooked by most policy-makers was the fact that "financial repression" not only kept interest rates artificially low but, by rationing credit, necessarily repressed the demand for certain types of credit (generally that for consumption). It was thus a serious oversimplification for neo-conservative theorists to focus exclusively on the favourable effects financial liberalization might have on effective savings and investment (via higher interest rates) and to neglect the unfavourable effect it could have on these by releasing the pent-up demand for consumption.[29]

8. None of this is to deny that financial repression has its cost, and that a move towards financial liberalization was in order. In retrospect, however, it seems clear that:
(a) Financial liberalization should not take place until after price stabilization

has been achieved or is well under way. The simultaneous pursuit of both jeopardizes the success of each, all the more so given the financial sector's sensitivity to disequilibria in other sectors, and its proneness to bubbles.

(b) Given the clear segmentation between international and domestic capital markets, and the further segmentation within the domestic capital market, financial liberalization should be pursued only as such segmentation is overcome or neutralized. Otherwise, the government must itself intermediate funds between the international and domestic markets, so as to control its flow and its cost, and determine who shall have access. Until segmentation is eliminated, some form of creidt allocation needs to be made, especially to assure adequate access to activities or sectors that have been neglected (small and medium-size firms $v.$ large firms, agricultural and construction $v.$ mining and manufacturing).

(c) Efforts should be made to assure the formation and operation of long-term capital markets before financial liberalization takes place; otherwise one risks moving most capital to the short terms, and distorting savings and investment behaviour.

(d) Given the economies of scale in finance and the temptation to form economic groups based on banks, and the subsequent leverage and distortions that may ensue, banking regulations need to be framed so as to limit bank–industry ownership links, to assure a wide distribution of ownership and control of banks, and to limit loans to any one single economic group or sector, especially if it is related to the bank itself.

(e) Upon liberalizing interest rates, efforts must be made to ensure that alternative mechanisms (e.g. excise taxes) are designed so as not to favour inadvertently a pent-up consumption demand, so that they will facilitate the achievement of the desired aim of raising ex-post national savings and investment.

Notes

1. Probably the most notable exponent of this position is R. McKinnon. See his *Money and Capital in Economic Development* (Brookings Institution, Washington, D.C., 1973) and "Represión financiera y el problema de la liberalización dentro de los paises menos desarrollados,"*Cuadernos de economía*, 47, (1979).
2. See, for example, V. Galbis, "Financial Intermediation and Economic Growth in Less Developed Countries: A Theoretical Approach," *Journal of Development Studies*, January 1977.
3. See R. McKinnon, "Represión financiera..." (note 1 above).
4. See chapter 11 of McKinnon's *Money and Capital...* (note 1 above), and J. Frenkel, "The Order of Economic Liberalization: Lessons from Chile and Argentina," in K. Brunner and A. Meltzer, eds., *Economic Policy in a World of Change*, vol. 17, Carnegie Rochester Series on Public Policy (North-Holland, Amsterdam, 1982).

5. For an extensive and detailed treatment of the process of financial liberalization in Argentina, Chile, and Uruguay, see R. Frenkel, "El desarrollo reciente del mercado de capitales en Argentina," *Desarrollo económico*, 78 (1980); J. Sourrouille and J. Lucangeli, *Política económica y proceso de desarrollo: La experiencia Argentina entre 1976 y 1981* (CEPAL, Santiago, 1983); R. Ffrench-Davis and J.P. Arellano, "Apertura financiera externa: La experiencia chilena en 1973–1980," *Estudios CIEPLAN*, 5 (1981); and I. Wonsewer and D. Saráchaga, "La apertura financiera," mimeo (Montevideo, 1980).
6. Each of the neo-conservative experiences can be usefully divided into two phases, in accordance with the focus of its price stabilization policy. In the first years, the attempts to bring down inflation centred on the pursuit of restrictive monetary, fiscal, and wage policies. Because these policies proved too slow and costly, in a second phase (beginning in 1976 in Chile and 1978 in Argentina and Uruguay) the price stabilization policy focused on controlling the exchange rate, letting monetary policy adjust passively, and thus hoped to force down domestic inflation to the rate of international inflation plus the pre-announced rate of devaluation via "the law of one price." This second phase, inspired by the monetary approach to the balance of payments, was to be especially decisive for capital inflows. For so long as there was confidence in the maintenance of the pre-announced exchange policy, capital could be especially sensitive to interest-rated differentials. And it is in fact in this second phase that capital inflows are especially strong. The third phase is the one after the maxi-devaluation and the abandonment of most of the neo-conservative policies.
7. Inasmuch as interest payments abroad strongly increased during the neo-conservative period, a better indicator of the domestic savings *effort* (or its restriction in consumption) would be national savings plus factor payments to the outside world, all expressed as a proportion of gross domestic income (this latter being GNP adjusted by the effect of variations in the terms of trade on national income). The domestic savings effort, so measured, as a proportion of gross domestic income is in Argentina slightly higher (1.5 percentage points) in the neo-conservative period than in the preceding 10 years; in Chile, it falls 3 percentage points in the same reference period; and in Uruguay it rises 3 percentage points.
8. Moreover, it is to be noted that because interest rates during the first half of the 1970s tended to be fixed and low, whereas in the second half of the decade loans were made at higher and variable interest, debt servicing costs in 1983 would prove to be higher than in 1975 even for similar debt to export ratios.
9. This may have been so (a) because of its greater spread for the foreign investor, at least in relation to Uruguay (which still fails to explain why the spread did not fall); or (b) because of its better growth prospects (in relation to Argentina).
10. Borrowing rates are the nominal rates of interest deflated by the wholesale price index, for this latter is all likelihood the most appropriate deflator for the debtor. Should these be deflated by the consumer price index, average real rates of interest remain unchanged in Chile, but fall from 17 to 5 per cent per year in Argentina and from 15 to 14 per cent in Uruguay. Interest paid to depositors was also high during the period, averaging 12 per cent per year in Chile, 1 per cent per year in Uruguay and −7 per cent per year in Argentina (negative, but far less so than in the past). Deposit rates naturally are deflated by the consumer price index, the more pertinent deflator for depositors.

11. This is suggested by an analysis of the Chilean case (see table 7) in which the cost of maintaining such non-interest-bearing reserves is estimated and in which, nevertheless, the average spread for the period remains close to 15 per cent. For a fuller treatment of this point see H. Cortes and L. Sjaastad, "El enfoque monetario de la balanza de pagos y las tasas de interés real in Chile," *Estudios de economía*, 11 (1978): and J.P. Arellano, "De la liberalización a la intervención. El mercado de capitales in Chile 1974–83," *Colección estudios de CIEPLAN*, 74 (1983).
12. See in this regard P. Spiller and E. Favaro, "An Economic Test of Interaction among Oligopolistic Firms: The Uruguayan Banking Sector," mimeo (Central Bank of Uruguay, Montevideo, 1982). Though there were 21 private banks, one state bank and many non-bank intermediaries, the authors argue that the legal barriers to further entry of new banks encouraged oligopolistic behaviour on the part of the existing financial system. It is striking, however, that spreads were similarly high in Argentina and Chile and behaved in much the same way, despite the absence of such barriers in the latter. This suggests that *in practice* it took a good deal of time for the pressure of competition to make itself felt.
13. Also important in explaining such high spreads may have been the fact that most operations were for 30 days, a fact which must have raised fixed costs. Yet this should have been largely compensated by the fact that there were a much greater number of operations than in the past, particularly since the periodic renovation of most short-run credits was effected almost automatically (and it was understood that this would be so), requiring little additional analysis.
14. In the absence of exchange risk and with no controls on capital flows, the nominal domestic interest rate should equal (or converge towards) the nominal international rate of interest plus the expected devaluation: $i_D = i_l + \bar{R}_e$. To the extent that the expected devaluation $\bar{R}_e$ is equal to the announced and executed devaluation $(\bar{R})$, $i_D = i_l + \bar{R}$. The exchange rate comes to lag inflation and be overvalued when $\bar{R} < \bar{P}_D - \bar{P}_l$. Therefore, when there is a lag in the exchange rate, $i_D < i_l + (\bar{P}_D - \bar{P}_l)$ or $(i_D - \bar{P}_D) < (i_l - \bar{P}_l)$. In other words, with an (expected) lag in the exchange rate, the real domestic interest rate should be less than the international rate of interest. And if the expected lag in the exchange rate were greater than the real international rate of interest, real domestic interest should be negative (or close to it, for there are country-risk and additional intermediation costs to be added in). Since LIBOR (6 per cent per year real) in this period was less than the expected lag in the exchange rate, substantially *negative* real domestic interest rates should have been observed for some of the period in question.
15. For theoretical treatments of some of the issues involved see T. Ho and A. Saunders, "A Catastrophe Model of Bank Failure," *Journal of Finance*, December 1980; J. Bullow and J. Shoven, "The Bankruptcy Decision," *Bell Journal of Economics*, autumn 1978; and F. Perez and A. Moreno, "Teoria financiera, contratos y políticas económicas," *Estudios publicos*, 14 (1984).
16. I owe this insight conerning the association of assymetry in the demand for credit and the degree of relative price (and so, wealth) changes to Carlos Massad.
17. Capital inflows are not simply and solely dependent on the differential in real interest rates between the domestic and international capital markets. In this period significant amounts of capital came in because the expected rate of return in dollars of direct investment was also high (at least as long as the exchange policy was ex-

pected to continue). This was the case with much capital brought in from overseas by nationals, not for investment in the domestic capital market but for making direct investments in the economy.
18. It is fascinating to consider not only *how* financial savings could go up without effective national savings and investment doing so, but also *where* these financial savings came from. On both these issues see J.P. Arellano (note 11 above). As to the latter question, he suggests, first, that 35 per cent of the increase in domestic financial assets was simply the counterpart of foreign debt, which was not spent on imports but augmented reserves till 1981. A second major source is simply capitalized interest rates on deposits, which between 1977 and 1981 alone were the equivalent of close to half of the existing non-monetary financial assets; thirdly, much represented a shift from government direct investment in public sector firms, funds which were increasingly channelled to the financial sector.
19. The *increased* share of foreign savings in GNP was estimated by comparing the average net increase in the annual flow of debt before and during the neo-conservative periods as a percentage of GNP.
20. For such "traditional" formulations see, among others, B. Sprinkel, *Money and Stock Prices* (Richard Irwin, Homewood, Ill., 1964); K. Homa and D. Jaffee, "The Supply of Money and Common Stock Prices," *Journal of Finance*, December 1971; and M. Hamburger and L. Kochin, "Money and Stock Prices: The Channels of Influence," *Journal of Finance*, May 1972.
21. See in this connection E. Fama, "Efficient Capital Markets: A Review of Theory and Empirical Work,"*Journal of Finance*, May 1970; J. Pesando "The Supply of Money and Common Stock Prices: Further Observations on the Econometric Evidence," *Journal of Finance*, June 1974; and C. Contador "Politica monetaria, inflação e mercado de acoes no Brasil-uma sintese de conclusões," *Revista brasileira de economia*, March 1974.
22. Of course, other factors were at work feeding such price speculation: the formation of economic conglomerates, of mutual funds, etc.
23. See P. Meller and A. Solimano, "El mercado de capitales chileno: laissez faire, inestabilidad financiera y burbujas especulativas," mimeo (1984), for an econometric test of the formation of a bubble (the speculative boom) and its subsequent bursting (the crash).
24. In Chile a multiple regression (corrected for second-order autocorrelation) between a real index of stock prices (RISP) and real M_1 (M_1 P) and real M_2 (M_2 P) and the real interest rate iR yielded:

$$RISP = -436 + 19M_1P - 2M_2P + 11 iR$$
$$t \quad (-6.9) \quad (9.4) \quad (-3.2) \quad (0.9)$$

with an adjusted R^2 of 88 per cent. The same regressions for Argentina had far weaker explanatory power (less than 5 per cent) and the interest rate was likewise not significant. The best result was achieved with $RISP_t = f(M_1P)_{t+1}$ but again iR was not correlated and had the wrong sign. The data are six-month moving averages for each quarter. Incidentally, since the stock price index and M_1P and M_2P are correlated in the same time period, these results could be fitted into the efficient capital markets model as framed by Richard Cooper, where stock market prices lend money (since money supply for the same time period would, in effect, be an expected value). See his "Efficient Capital Markets and the Quantity Theory of Money," *Journal of Finance*, June 1974.

25. The quarterly percentage variation in RISP (VRISP) regressed on the quarterly percentage variation in real $M_1(VM_1P)$, real $M_2(VM_2P)$ and in the absolute change in iR(ViR), and corrected for autocorrelation, showed no significant correlation with interest-rate variations in time t. But in Chile $VRISP_t$ was negatively correlated with ViR_{t+1} (with 90 per cent confidence) and positively correlated with ViR_{t-1} (with 99 per cent confidence). In Argentina $(VRISP)_t$ was correlated with VM_2P (with 89 per cent confidence), lending some additional weight to the hypothesis of Richard Cooper that stock-market price variations may actually lead money supply changes. See his "Efficient Capital Markets..." (note 24 above).

26. For a detailed treatment of the financial crisis in the Southern Cone and alternative ways of dealing with it, see among others, E. Barandiarán, "Nuestra crisis financiera," *Estudios públicos*, 12 (1983); Carlos Díaz-Alejandro, "Goodbye Financial Repression, Hello Financial Crash," mimeo (New Haven, Conn., 1983); and R. Fernández, "La crisis financiera argentina: 1980–1982" (CEMA, Buenos Aires, 1982). A somewhat more upbeat interpretation of the prelude to the crisis in Chile can be found in D. Mathieson, "Estimating Models of Financial Market Behavior during Periods of Extensive Structural Reform: The Experience of Chile," *IMF Staff Papers*, June 1983.

27. Of course, how best to overcome such disequilibria is another question. Intervene in the money market, the goods market or the asset market? Expand the quantity of money, validate the prevailing level of prices, and thus avoid a recession or control interest rates directly? Prohibit the renewal of credits without adequate guarantees or put an end to loans to firms in the same conglomerate as the bank and thereby speed the liquidation of assets? These issues, though important, go beyond the purposes of this study.

28. This was especially the case where financial opening up was more limited (Chile). See R. Zahler, "Repercusiones monetarias y reales de la apertura financiera al exterior: el caso chileno, 1975–1978," *CEPAL Review*, 10 (1980); the author estimates that such segmentation implied a transfer of the order of US$1 billion to those firms that enjoyed privileged access to foreign credits.

29. Though a case can be made that consumer durables are a form of savings, and certainly may improve welfare, the point is that the hoped-for increase in *productive* investment was thereby squelched.

THE FAILURES OF THE CAPITAL MARKET: A LATIN AMERICAN VIEW

Eduardo Sarmiento

Introduction

In the classical theory of capital markets, competition leads to the most efficient use of resources and government intervention introduces distortions and rigidities that interfere with this process. Thus, the financial liberalization of economies that have been subject to controls for several years results in greater saving, the development of more efficient activities, and a rise in the growth rate.

This classical conception is based on an ideal market. It assumes that economies have great capacity for adjustment, in the sense that goods are easily substitutable, supplies relatively elastic, prices highly flexible, and markets made up of several individuals. However, some underdeveloped economies have certain features that depart considerably from these assumptions. Are such features significant enough to modify the results of the classical model and to explain the failure of the capital market? An attempt will be made to answer this question on the basis of some central theoretical concepts and on the experience of financial liberalization in Latin America.

The article is made up of three parts. The first identifies certain special features of developing economies that have not been included explicitly in the analysis of capital markets. Complementarities between credit and demand for goods, supply of physical assets, monopolistic power of economic agents, and the prevalence of adaptative expectations are examined in conceptual terms. Since these elements are irregular in the sense that they do not generate smooth adjustments or lead to Walrasian equilibrium situations, their implications for capital markets cannot be analysed by traditional approaches. Some concepts and tools are suggested for managing these kinds of problems.

The second part is concerned with financial liberalization. It does not include a detailed description of policies and institutional changes, but is an attempt to analyse empirically the liberalization of certain Latin American capital markets. The information generated by the reforms provides information for the evaluation of the response of certain variables to policy changes and to test different hypotheses.

Part 3 attempts to sum up the degree to which the information of part 2 provides evidence of the features of capital markets introduced in part 1. Furthermore, these elements are considered simultaneously to summarize the causes of the failures of capital market and to discuss market liberalization. Finally, the financial reforms required to reconcile efficiency and stability are outlined.

The analysis is based on empirical evidence from Colombia. Since many arguments are applicable to other developing countries, the results obtained from Colombia are also illustrated in some cases with information on Argentina and Chile.

1. Basic Concepts

Credit and Durable Goods

There are some empirical studies that suggest that the credit effects vary with the characteristics of goods. The evidence is especially clear for durable and non-durable goods. This result can be interpreted in terms of the theory that distinguishes flow and stock variables. It is generally recognized that flows are related to income and stocks to wealth, and they are determined in the short run by a different set of elements. Under this framework, credit fluctuations initially affect stocks and wealth and only later non-durable goods. Relative prices of the two kinds of goods may fluctuate in the short run, interfering with the adjustment process.

These concepts can be illustrated with the wealth identity:

$$W + C = P_K K + P_D D + F$$

where W = wealth, C = credit, K = capital goods, D = durable goods, P_K and P_D = prices of capital of consumer durable goods, and F = financial assets.

According to this identity, the sum of wealth and credit is equal to physical and financial assets. If credit increases and production does not vary, the prices of investment and durable consumer goods will rise. The funds come back to the financial market in the form of financial assets or they are used to acquire physical assets. The effective wealth of the economy becomes greater than desired wealth. Equilibrium is maintained through a price level rise or if credit contracts. This relation also holds in the opposite direction. The stock price increase can be effective only when desired wealth increases or credit expands.

For this reason the demand for credit increases in the same direction as asset prices.

Credit expansion determines, then, an increase in the relative prices of physical assets and, later, an increase in financial assets. Both factors jointly determine an expansion of effective wealth of the economy with respect to desired wealth. As a consequence, individuals will substitute physical assets for non-durable goods, generating a rise in the general price index. The new equilibrium will be reached when relative prices return to their initial levels.

This adjustment does not occur when prices of non-durable goods are subject to controls or are determined by external prices in an open economy. The system will remain in disequilibrium unless asset prices return to their original point. But asset owners will interfere with the price decrease, trying to hold the assets and withstand the pressure to sell in order to maintain the initial increase of credit. Therefore a general tendency to avoid asset liquidation at a lower price may produce a high demand for credit, preventing a decline in the interest rate.

Usually, it is supposed that asset prices are determined in a system in which demand and supply are satisfied. However, in the real world economic agents do not have the information that tells them when the economy is in this situation. For instance, in stock markets, which are subject to imperfections, prices do not always reflect effective transactions. It is common in these markets that individuals fail to sell their products because they offer them at prices above competitive levels. It is possible that under a system of adaptive expectations, in which individuals tend to fix prices according to their previous experience, perceived prices will turn out to be higher than those that satisfy the wealth equation. This is a typical situation in which perceived wealth is greater than effective wealth. If individuals try to hold these prices, the demand for credit will be higher than the amount of credit that causes the initial shock. Interest rates will be determined by unreal price expectations and, therefore, may settle at any level.

It is clear that physical assets and credit are closely related; that is, they are complementary goods. Although this fact is not rejected openly by economists, it is not always incorporated in practice. The influence of the neo-classical theory has led to the assumption that all goods are substitutes. Under these conditions, the fluctuations of one market, although influencing other markets, do not change their main features. The formulations in which the conditions of supply, demand, and interest rate of the financial market are independent of the rest of the economic system can be justified. But this simplification is misleading when goods are complements. In such cases, the fluctuation of one market can destabilize another that, in isolation, works properly. Thus, the adjustment described previously shows that credit affects asset prices and they affect interest rates. This link between the credit market and the physical assets market will play an important role in the interpretation of real life. It will be shown that many events that occur in the credit market originate in the physical assets market, and vice versa.

The final impact of credit depends on supply conditions which cannot be generalized. The adjustment process described in this section mainly applies to durable goods with inelastic supply. This is the case with assets that require a long time for production, such as housing and assets that are subject to indivisibilities.

Liquidity and Indivisibilities

The possibility of buying a good and selling it later at a higher price is limited by the resources that are required to carry out the transaction. Thus, liquidity emerges as one of the main requirements to accomplish any speculative transaction and it is closely related to the characteristics of goods. If goods were divisible, speculative options would be very scarce. Everybody may participate in these operations, as is the case in the non-durable goods and service markets. The expected value of speculative earnings tends in the long run to be equal to the return of capital. Since fewer people can participate in these markets, the greater are the indivisibilities; speculation in markets of goods of great size gives monopolistic power to individuals that have access to liquidity. This is verified by actions applied by economic groups in the transactions of major enterprises. They induce price changes that allow them to obtain great profits in buying and selling operations.

The Keynesian concept of speculation refers to the fact that individuals hold liquid assets, waiting for the opportunity to earn profits through transactions in goods. Indivisibilities modify this notion in the sense that liquidity is required not only to take advantage of an opportunity but also to acquire a good to which others do not have access. Profits do not come from taking advantage of buying a stock at low cost, but rather from the administration of a business that requires liquidity.

At this point it is important to distinguish property with control from property without it. The latter refers to small stockholders and the second to economic agents that manage and guide enterprises. While the attitude of small stockholders is influenced by short-run elements, the goal of stockholders with control is determined by the real situation of enterprises and its effect on other related activities. Under these conditions the hypothesis that major enterprises become divisible through the stock market is not true. Property is indivisible, because a high share of assets is required in order to get real control.

It is generally assumed that physical assets and liquidity are substitutes, in the sense that individuals can acquire more goods to the extent that they reduce liquid assets. Our previous arguments suggest that this is not the case. Some goods are complementary with liquidity and the relation between the two rises with indivisibilities. The demand for those goods tends to be restricted by quantity. Liquidity appears then as a restriction in the operation of the economic system that prevents individuals from satisfying their desires.

In economic theory, it is usually supposed that distortions that obstruct market adjustment take place in an isolated and moderate way. However, this interpretation cannot be applied easily to liquidity, which influences many individuals and goods. The limitations originated by the control of large enterprises result in an effective financial asset demand inferior to the desires of economic agents, which has deep implications for the way the economic system works.

Can this restriction be corrected by increasing the quantity of money? The answer is provided by the concepts developed in the previous section and it is related to the characteristics of major enterprises. The construction of large enterprises takes many years, and so, in the medium term, the supply of these enterprises is inelastic. The increase in the demand to acquire them manifests itself initially in an asset price increase. Since this effect does not last long, the real quantity of enterprises remains unchanged. The increases of the relative prices of assets determine an increase of wealth, stimulating expenditure expansion and a rise of the general price level. In this way, the initial increase of asset prices and the expansion of the nominal quantity of money are neutralized. Relative prices and the real quantity of money tend to return to the initial point. A vicious circle is created. Output does not increase because the price increase is transitory; it is transitory because output is unchanged.

Substitution between Physical and Financial Assets and the Stock Market

There has been a long debate about the substitution between physical and financial assets[1]; Sarmiento[2] shows that this notion is not valid in global terms. The shift of wealth towards financial assets allows for the support of the same quantity of physical assets with less equity. The interesting relation is between financial assets and physical capital, and it depends on the way that credit is generated and financial expansion is oriented. If it is used to finance durable consumption goods or to maintain speculative balances, financial and capital assets appear as substitutes.

These concepts are more theoretically meaningful and have a more practical use when they refer to specific economic agents. Substitution between financial and physical capital is different for regular stockholders than it is for holders with enterprise control. It is likely that the former are willing to exchange stock for financial titles if they earn a higher return. In contrast, the return of stockholders with control is determined by the long-run conditions of enterprises, and that cannot be easily swayed by current yields on financial titles. Furthermore, it seems that enterprise control transactions are limited by liquidity. The buyer of an enterprise should be in a position not only to make a patrimonial transfer, but also to maintain the credit to finance a high share of assets.

In sum, for regular stockholders physical assets are highly substitutable for financial assets, while for stockholders with control they are relatively complementary.

The link between credit markets and stock markets appears clear within this framework. Suppose that the interest rate of financial titles rises through administrative actions or through market forces. Regular stockholders will shift to financial titles and this will increase credit. If enterprises are able to attract resources, they can face a situation in which their liabilities are represented more in credits and less in equities. In global terms, financial assets and credit grow, the number of regular stockholders decreases, and the economic power of stockholders with control expands.

The different motivations of regular stockholders and stockholders with control have had a central influence in the behaviour of stock markets and financial instruments. Some of the failures observed in recent years originated in misperception of this fact. The hypothesis that the two kinds of stockholders behave in the same way led to the assumption that financial title expansion came from a net growth of savings, and the weakness of the stock market did not affect investment and economic power. These assumptions had a decisive influence on the government's passive attitude towards the irregular performance of the financial sector.

Stock-market evolution in the last 25 years in Colombia provides valuable information about these phenomena and allows the testing of some hypotheses. Figures are based on a sample of 30 enterprises collected by Comisión de Valores and published in a paper by J.C. Restrepo et al.; the basic information of the sample is summarized in tables 1 and 2. The first contains the consolidated balance of enterprises, and the second shows the average behaviour of distributed profits and stock prices.

In the Appendix, an approximated expression for the expected return of stock is derived. This expression is equivalent to the interest rate produced by a financial title with the same risk. Thus, in equilibrium, the financial market and the stock market are related by the following equations:

$$r = \frac{D}{Pa} + \frac{\dot{D}}{D} - \frac{\dot{P}}{P},$$

where D = the dividend, Pa = stock price, P = price level, $\dot{P}/P$ = rate of change of the price level, $\dot{D}/D$ = rate of change of dividends, and r = the real interest rate.

Dividend and inflation data shown in table 2 were introduced in the previous equation to estimate the equivalent interest rate. The results of the exercise are summarized in table 3.

Tables 1 and 2 show that stock prices have decreased in the last 20 years.

Table 1. Financial structure of corporations in Colombia, 1960–1981 (percentages)

	1960	1961	1962	1963	1964	1965	1966	1967	1968	1969	1970	1971	1972	1973	1974	1975	1976	1977	1978	1979	1980	1981
1. Assets																						
Cash	4.0	4.0	5.0	4.0	4.0	4.0	4.0	3.0	3.0	3.0	2.4	2.5	2.6	3.1	2.9	3.1	3.2	3.4	3.9	3.2	3.9	3.3
Debtors																						
Short-term	20.0	20.0	21.0	23.0	24.0	25.0	26.0	24.0	24.0	25.0	23.7	23.7	24.3	25.8	23.8	25.1	26.8	28.3	26.3	27.5	29.5	27.6
Long-term				1.0	1.0	1.0	2.0	2.0	2.0	3.0	3.1	2.7	2.9	3.1	3.4	4.4	4.3	3.9	4.3	5.7	4.0	3.1
Inventories and inv.	31.0	29.0	27.0	26.0	26.0	25.0	28.0	26.0	24.0	27.0	26.4	27.2	26.1	25.5	30.3	30.3	30.9	30.5	30.1	29.7	30.0	24.1
Plant and equipment	31.0	32.0	33.0	31.0	29.0	28.0	26.0	28.0	28.0	28.0	21.9	21.6	21.5	21.2	19.2	19.6	18.8	17.4	15.9	15.1	20.8	29.8
Other assets	14.0	15.0	14.0	15.0	16.0	17.0	15.0	17.0	19.0	14.0	22.5	22.3	22.6	21.3	20.4	17.5	16.0	16.5	19.5	18.8	11.8	12.1
Total	100	100	100	100	100	100	100	100	100	100	100	100	100	100	100	100	100	100	100	100	100	100
2. Liabilities																						
Creditors																						
Short-term	31.0	32.0	35.0	26.0	28.0	25.0	29.0	26.0	26.0	27.0	26.5	27.4	28.1	30.9	34.3	34.1	34.6	38.3	37.3	40.3	44.3	41.0
Long-term	6.0	5.0	5.0	8.0	10.0	13.0	12.0	14.0	14.0	15.0	10.1	12.2	14.3	13.6	13.0	16.0	15.0	12.3	14.1	14.8	16.7	22.2
Other liabilities				6.0	5.0	5.0	4.0	4.0	4.0	4.0	7.3	7.8	7.4	9.2	9.7	10.4	12.5	11.0	9.6	10.0	8.6	8.5
Total liabilities	37.0	37.0	40.0	40.0	43.0	43.0	45.0	44.0	44.0	46.0	43.9	47.4	49.8	53.7	57.0	60.5	62.1	61.6	61.0	65.1	69.6	71.7
Equity	31.0	31.0	30.0	28.0	27.0	26.0	24.0	23.0	22.0	21.0	16.1	14.4	12.8	11.2	9.5	8.3	7.8	6.6	5.4	4.8	7.1	9.0
Profits	6.0	5.0	6.0	6.0	5.0	4.0	5.0	4.0	3.0	5.0	6.6	6.0	6.1	6.3	6.8	5.4	6.3	6.5	7.8	6.1	5.3	3.9
Surplus	26.0	27.0	24.0	26.0	25.0	27.0	26.0	29.0	31.0	28.0	33.4	32.2	31.3	28.8	26.7	25.8	23.8	25.3	25.8	24.0	18.0	15.4
Total	100	100	100	100	100	100	100	100	100	100	100	100	100	100	100	100	100	100	100	100	100	100

Source: Camilo Pieschacón, "La política económica y el mercado bursátil," *El mercado de capital en Colombia, ahorro y crédito 1973* (Banco de la República, Bogotá, 1974). Calculations of the National Commission of Values, Division of Market Development, Bogotá.

Table 2. Dividends, quotations, and expected return of industrial shares in Colombia

Year	Dividends of industrial shares (%)	Quotation (index)	Expected return (%)[a]
1960		178.0	
1961	14.5	163.0	
1962	12.0	160.5	−2.3
1963	11.7	131.7	−8.3
1964	12.5	125.8	11.1
1965	13.4	106.4	−0.6
1966	15.2	83.6	20.6
1967	15.2	83.3	20.7
1968	13.8	89.4	11.0
1969	10.9	97.8	3.1
1970	10.0	100.0	9.7
1971	14.7	75.5	14.3
1972	16.4	55.5	11.1
1973	17.1	46.1	3.0
1974	14.2	33.4	10.4
1975	22.0	83.6	8.5
1976	18.7	23.6	8.6
1977	17.8	26.4	19.9
1978	10.8	33.3	3.9
1979	9.6	33.9	9.9
1980	15.1	23.6	6.0
1981	16.0	23.0	−37.5
1982		22.8	

a. See Appendix for definition: $r = \dfrac{D}{Pa} + \dfrac{Pa}{Pa} - \dfrac{P}{P}$

Source: J.C. Restrepo, J.G. Serna, and M.G. Rosas, "Inflación, financiamiento y capitalización empresarial," *Ensayos sobre política económica* (Banco de la República, Bogotá, 1983).

However, the trend has not been a smooth one. Prices fall in some periods and rise moderately in others. Furthermore, the pattern has changed over the last two decades.

In the decade of the 1960s it is possible to distinguish two periods: stock prices fell from 1960 to 1966 and rose between 1967 and 1970. It is also observable that, in the last period, dividends were higher and grew at a higher rate than inflation. Thus, expected return was 4 per cent from 1960 to 1966 and 11 per cent from 1967 to 1970. The close relation between real prices and expected returns is the evidence that the stock market was highly sensitive to short-term returns. This market was dominated by economic agents for which stocks and financial titles were close substitutes.

Relations are not so clear in the decade of the 1970s. The huge fall in stock

Table 3. Equivalent interest rates and real quotations in Colombia

Period	Equivalent interest rates (annual %)	Average variation of real prices (annual %)
1962–1966	4.02	−12.1
1967–1970	11.1	−4.6
1971–1975	10.5	−24.9
1976–1979	10.6	10.0
1980–1982	−15.7	−10.9

Source: Estimates of the author.

prices between 1970 and 1975 cannot be attributed to dividend behaviour. The expected rate of return in that period was not significantly different from that of 1966–1970. The same may be said about the price recovery that took place in the period 1977–1979, when expected returns did not show any major difference with respect to those of previous years. These factors suggest that stock-market behaviour during the 1970s cannot easily be attributed to short-run phenomena. The explanation can be found in structural changes in the stock markets and in external factors.

Table 3 shows that the respective returns were higher in the 1970s than in the 1960s. This result is related to the transformation of the financial titles during the 1970s. The increase of the interest rate on financial titles during this period led to competition that induced individuals to ask for higher dividends to hold stock.

Fluctuations between 1970 and 1980 reflect several facts accumulated before and during the financial transformation. Since the process of price reductions and shifts of regular stockholders to financial titles reduced the share of equity in liabilities, the control of enterprises could be obtained with less patrimony and produced large profits. The influence of stockholders with control grew fast and substitution between stock and financial titles weakened progressively. The close relation between prices and expected return observed during the 1960s disappeared. Thus, the rise of stock prices in 1977–1979 was not stimulated by an improvement in returns, but by competition between economic agents to obtain control of enterprises. This is why price fluctuations increase and become more random in the short run.

The stock market's weakness cannot be attributed, however, to inevitable causes. High substitution between physical and financial assets for regular asset holders and low substitutability for stockholders with control is probably the only factor that can be considered inherent to the working of the economic system, but it could not generate the transformation by itself. The fall of stock prices, the shift of regular stockholders towards financial titles, and the property concentrations are related to factors that appeared and were strengthened throughout the two decades. For the 1960s it can be explained by low returns of stocks, and in the 1970s by the rise of interest rates and the increasing power of economic groups. But nothing was inevitable. The low returns of the 1960s could have been prevented by a wider earnings distribution policy. The rise of interest rates, which had their origins in the financial liberalization policy, could have been neutralized through administrative procedures. It is true that, at the end of the decade, the power of groups could be opposed only at the expense of a serious crisis. But it is also true that this situation would not have existed without the conditions that facilitated the process of concentration of economic power.

Tax legislation is another factor that has affected the stock market. The prevailing regime in the last years includes norms of different kinds that discriminate against stocks. The most important among them is the one that allows enterprises to charge interest payments as costs. Since these payments have an inflation component, which corresponds to valorization of liabilities, credit has a subsidy that grows with inflation and hides real profits. This framework stimulated the substitution of capital equity for credit, allowed a policy of low distributed earnings, and contributed to an inequitable system.

These results have practical and theoretical implications. There is no doubt that economic groups have monopolistic power that allows them to influence stock prices. The policy of increasing nominal dividend at a lower rate than inflation induced people to sell stocks, leading to the fall in their prices. Of course, this behaviour was favoured by the neutral attitude of governments. The actions of the economic groups were facilitated by the increase in the interest rate of financial titles and by the lack of effective regulations.

This behaviour of the stock market cannot be attributed to the special conditions of the Colombian economy. Figure 1 shows the evolution of stock prices in Colombia, Argentina and Chile. The behaviour is similar in the three countries. Real stock prices fell for a long period, recovered around 1975 and fell again around 1980. The explanation for Argentina and Chile is not different from the one given previously for Colombia. The fall that took place before 1975 was caused by low returns of stock and by the lack of an appropriate secondary market. The recovery that starts in 1975 was generated by the actions of economic groups to take control of the enterprises by buying up the equity of regular stockholders. This monopolistic competition brought about a huge rise of stock prices, originating expectations that contributed to their settling at a level at which dividends were way below interest rates. Situations of bankruptcy and insolvency emerged gradually among the economic groups that had bought such stock on credit and led to a general collapse, which, among its multiple results, generated an enormous price fall.

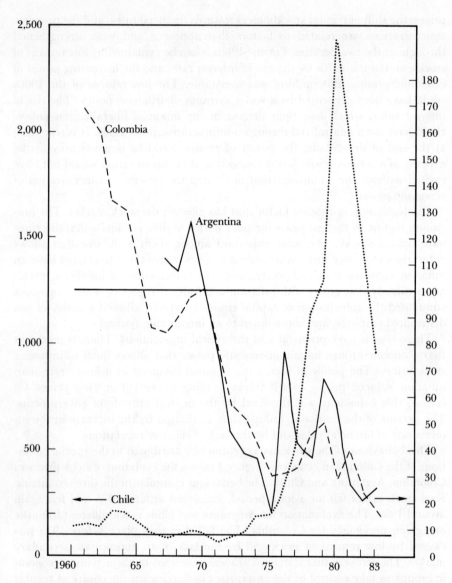

Fig. 1. Real stock price index in Argentina, Colombia, and Chile (1970 = 100)

Economic Groups

Industrial and financial development in Latin America in the last decade has been associated with economic groups or conglomerates. These groups share some characteristics of the German and American models. They are made up of individuals that have similar links and profess great loyalty to the organization. Groups operate in the financial and the industrial sector and, in general, attempt to expand to multiple activities. Their basic objective is to get maximum benefit for the organization, which in some cases means that certain activities generate losses in order to get profits in others.

One of the outstanding characteristics of groups is their attraction to high-risk activities. However, the usual explanation that groups are in a better position to diversify risk, because they operate in different activities, is not sufficiently convincing. There are more powerful reasons.

The most important perhaps is related to the activities of groups. Some of these activities have the properties of public goods and generate externalities for other activities. This is the case with financial institutions, whose failure brings about a loss of confidence that affects the stability of the financial system and of the whole economy. A similar phenomenon occurs with major enterprises, whose health is interpreted as an indicator of the state of the whole sector and as a responsibility of governments. The bankruptcy of these enterprises reduces employment, since machines cannot be moved from one activity to another. This can be prevented by government subsidies, which usually correspond to a small part of fixed costs. In the short run, the private and social costs of maintaining working plants are less than the cost of closing them. This is the reason why in practice no government assumes under its administration the dissolution of major enterprises.

On the other hand, groups are organized in such a way that profits in different activities are not associated. The owners of the enterprises are registered with different names and the practice of generating losses in some activities to get profits in another is highly generalized. Simultaneously, actions of groups in related activities that do not have shared responsibility leads to behaviour that cannot be understood in terms of traditional concepts.

Groups that operate within this framework obtain the power to administer a public good. They are in a position to appropriate the profits of activities that affect a large proportion of the population. As a result, their efforts are aimed at concentrating the profits generated in the good times on group owners and at transferring the losses of bad times to governments and regular stockholders.

This is a typical game of large positive elements.[3] The expected profits of groups are substantially greater than zero and increase with the number of activities. There are great incentives to expansion. The only limitation on group actions is credit and liquidity and seldom the administrative diseconomies that result from disparate activities.

The expansion capacity of groups leads one to assume that they have great potential to accomplish new high-risk investments. But the attitude depends on the characteristics of the capital market. If the returns on buying existing enterprises were higher than the returns of new investment, obviously groups would move to take advantage of this option. The organization would concentrate on operating quickly and with large liquidity in several activities. Profits would come more from speculation and transfer activities than from wealth creation.

Group organizations depend on economic and institutional conditions. In an economy with great speculative options and no limitations on the actions of economic agents, they will try to form efficient systems to manage liquidity and to obtain the advantages of a large positive elements game. The profits in such a system will not come, certainly, from market competition, but from its weaknesses.

Heterogeneity of Capital Markets

One of the weaknesses of economic theory lies in the assumption that all markets behave like the market for homogeneous and perishable goods. The adjustment of these markets takes place in a short time. Producers assume that the rise of the relative price of their products is transitory, or, at least, that the earnings produced by storing them are small. The supply of and demand for different products are always in equilibrium and expectations are formed on the grounds that the interest rate will not change. In contrast, asset markets, because of imperfection and low supply response, have slow adjustments. Investors, who know very well that these prices are subject to cycles that could last many years, interpret price fluctuations as something that will continue in the future. A speculative demand occurs, accentuating price movements. Thus, the price rise of stocks and urban property generates expectations that tend to reinforce it.

This error is common in the theoretical formulation of the financial market. As in the case of homogeneous products, it is assumed that the market is always in equilibrium and that the interest rate reflects resource scarcity. We have seen, however, that physical assets are highly complementary with credit and that the fluctuations of the market for physical assets move inevitably to the financial markets. The adjustment problem of one market is reflected in the other. In fact, the tendency of individuals to suppose that price increases will continue in the future leads to an increase in goods and credit demand, which finally determines a rise in the interest rate. This result is contrary to orthodoxy. Physical asset prices and interest rates move in the same direction.

The presence of economic groups has contributed also to a modification of the characteristics of adjustment. These groups develop within a large positive elements game which allows them to obtain losses in certain activities for some time. Profits come in the long term and, sometimes, are generated when transactions take place. Therefore, group decisions are not very sensitive to transitory price changes and they tend to finance temporary losses with credit expansion. Under those conditions, asset prices and interest rates rise and the demand for credit increases. This behaviour is facilitated and accentuated by the link between financial and industrial activities. There are no filters or banking guarantees that condition financing of enterprises to its capacity to serve the debt. It is well known that sometimes groups make use of banks in order to extend indefinitely the loans of their enterprises.

It is evident that the characteristics of financial markets within a framework in which credit is utilized to obtain monopolistic gains are different from the ones of the competitive markets described in textbooks. It is usually assumed that there are different kinds of forces that regulate and stabilize the market. In contrast, in monopolistic markets there could be a permanent pressure, in both expansion periods as well as contraction periods, to hold and expand the stock of indebtedness. The demand for credit would be infinite, or at least greater than supply, leading to high interest rates and introducing instability in the system. There is nothing that prevents and regulates the stock price and the interest rate rise. The process can take place for a long time, driving the economic system to collapse.

The infinite demand for credit, of course, is a short-run phenomenon. The rise of asset prices and the interest rate lead progressively to insolvency. The power of groups might prolong this process for many years, but cannot avoid the collapse that usually materializes in the bankruptcy of banks and enterprises and great losses by savers. The price and interest rate fall occurs abruptly.

2. Financial Liberalization

During the decade of the 1960s the Colombian economy developed within a highly regulated financial system. Credit was oriented preferentially to enterprise investment, interest rate and financial title characteristics were determined by monetary authorities, and external credit was centrally regulated and limited for certain kinds of projects. The liberalization comprises two kinds of policies. On the one hand, banks were authorized to negotiate the interest rate on loans, to fix asset interest rates above inflation, and to determine the characteristics of financial instruments. On the other hand, banks were permitted to orient a greater part of their portfolios freely, the role of orienting special priority activities was reduced, and the tendency to avoid credit concentration was weakened.

However, the Colombian financial liberalization policies were not like the ones followed in other Latin American countries. For instance, Argentina and Chile went much further, especially with regard to credit. In both countries, banks were authorized to distribute freely their portfolios and restrictions on external financing were eliminated. In contrast, in Colombia, the liberalization of internal credit was partial and external credit was maintained under central government regulation.

Credit Liberalization and Financial Instruments

Before the liberalization, the Colombian economy operated within a system of regulated interest rates and selective credit. The resources were oriented preferentially to enterprise investment. The other activities, loans for durable consumer goods, and speculative activities, were rationed and, as such, consumers were willing to pay an interest rate higher than the one fixed by the monetary authorities.

Credit liberalization provoked, of course, a movement of resources towards previously restricted activities. This process was especially visible in durable consumer goods, whose demand increased notoriously. The price of housing rose and the expenditures on automobiles and home appliances expanded significantly. Besides, the process of interest-rate rise started in these areas. The practice of asking for and paying excessive interest rates began with the loans for automobiles and home appliances.

The most interesting case is the one of the large enterprises. The traditional restrictions on the use of credit to acquire existing activities and to avoid the portfolio concentration of intermediaries depressed the prices of many of those enterprises. This situation was a result of the complementarity between liquidity and the control of big enterprises. Credit liberalization opened up the opportunity to buy them at below their real value. But the option was restricted to groups that had access to financial resources and the organization to manage these enterprises. The practice of acquiring enterprises through credit became an attractive business and it soon generalized.

The process originated a strong link between industrial activities and the financial sector. The groups purchased enterprises with credit given by their bank and created conditions of monopolistic competition. The indebtedness of enterprises increased, stock prices rose abruptly due to take-over activities and the financial sector became more vulnerable.

The rise in stock prices generated by financial liberalization was a generalized phenomenon. Table 4 shows the evolution of these prices for Colombia, Argentina, and Chile in the decade of the 1970s. The periods before and after liberalization cannot be established precisely because policies were undertaken over periods of several years. Tentatively, it can be assumed that liberalization was realized in 1973–1975 in Colombia, in 1974–1975 in Chile, and in 1976–1977 in Argentina. The stock price response was different in the three countries, because the characteristics and expectations of liberalization policies were also different. However, in all three countries liberalization was followed, with different lags, by enormous stock price increases. The rise occurred in Argentina in 1975–1977, in Colombia in 1976–1979, and in Chile in 1975–1980. However, the intensity of the phenomenon was different in the three countries. It was monumental in Chile, significant in Argentina, and moderate in Colombia.

The fall in prices and their subsequent rise cannot be interpreted using the

Table 4. Real stock price index in Argentina, Colombia, and Chile (1970 = 100)

Year	Argentina	Colombia	Chile
1960			
1961		167.8	135.5
1962		161.1	136.6
1963		134.1	205.9
1964		129.9	191.9
1965		109.4	148.6
1966		86.4	98.9
1967	112.9	84.3	87.3
1968	106.9	89.5	88.0
1969	125.6	97.4	111.2
1970	100.0	100.0	100.0
1971	80.9	76.1	61.0
1972	48.3	57.8	85.5
1973	39.0	52.0	98.3
1974	37.1	41.9	180.5
1975	17.1	30.2	189.7
1976	76.7	30.8	313.1
1977	43.6	33.3	576.6
1978	36.9	45.6	1,141.8
1979	66.5	50.4	1,326.0
1980	56.8	28.8	2,443.8
1981	27.0	39.8	1,963.2
1982	20.3	25.1	1,486.2
1983	25.5	16.2	966.9

Sources: Commercial Market of Buenos Aires; Stock Market of Bogotá; Commercial Market of Santiago de Chile; and National Institutes of Statistics of Argentina and Chile.

traditional concepts of perfect competition. There is no doubt that stocks, after being undervalued as a consequence of credit restriction, become overvalued. This change can be understood only when the optimistic expectations that accompany financial liberalization are recognized. The rise in stock prices validated the assumption that before they were undervalued, inducing additional movements in the same direction.

The liberalization of financial paper was also very influential. In Sarmiento[4] it is shown that in the 1960s the financial opportunities were limited to stocks, savings deposits, and some titles with fixed interest rates. Financial reforms modified this framework, authorizing financial intermediaries to issue titles at any term and to fix freely the interest rate of many of them. At the same time, intermediaries were permitted to negotiate with clients the interest rates of loans. The supporters of financial liberalization considered that the discretionality of financial intermediaries would be regulated efficiently by the market, which would work as a filter, selecting the best options for society.

These hypotheses disregard the role of liquidity. Considerable evidence of its importance is found in the Colombian economy. Estimates of the substitution between financial assets suggest that individuals require great incentives to shift to longer-term bonds. It seems that this maturity is not determined by the return of resources in productive activities, but in speculative activities. This argument is confirmed by events that occurred during the liberalization process. In Sarmiento it is shown that the issuance of short-term financial assets displaced all long-term financial assets. Because of liquidity preferences, individuals hold the latter only if they yield much more than the former. Thus, the interest rate is too high and the real sector cannot afford it. Enterprises, to avoid the shift from stock to financial titles, had to raise the returns of stocks by raising dividends above the market interest rate, something that could not be done, given the level of normal industrial profits. Of course, this behaviour was aggravated because the economic groups were not interested in having independent stockholders continue to hold stocks. In fact, they followed a restrictive policy of profit distribution that accelerated the fall in stock prices and increased the public's shift from stock to financial titles.

These changes altered the financial structure of enterprises. They moved aggressively to substitute equity for loans of the banking system. But this transformation did not depend only on them. Short-term resources provided by savers could not be transferred directly to enterprises that required long-term loans to accomplish long-run activities: the link required the support of financial intermediaries. By the law of large numbers, withdrawals and deposits from banks do not occur simultaneously, which places them in the position of receiving short-term funds and lending them for a long term. In this way enterprises have access to loans with terms greater than the ones of average deposits or highly liquid financial instruments.

Clearly, credit dependence led to a strong link between the industrial sector and the financial system. The stability of enterprises depends on the willingness of banks to turn liquid deposits into medium-term loans. Moreover, the control of enterprises is easier when liabilities are represented more in credit than in equity. Fewer real assets can be obtained with small amounts of one's own capital. Credit appears then as a critical determinant of financial market organization and access to it provides monopolistic power.

Financial and Industrial Linkages

Financial liberalization created great possibilities for using credit to organize the take-over of industrial and financial enterprises. However, these opportunities were restricted to organizations that had the means to administer great enterprises and to operate multiple activities simultaneously. For this reason, financial liberalization coincides with the strengthening and expansion of economic groups.

The strengthening of economic groups resulted from the monopolistic power given by access to credit and from the advantage of the financial and industrial link. The most visible efforts of groups were oriented to operations that required high liquidity and quick decisions and to the purchase of existing enterprises. The expansion of these groups was reflected more in the price of existing enterprises than in the quantum of new investment, and did not generate more efficient activities. The balance sheets of the ten largest enterprises indicate that their profits have decreased in the last five years and that many of them are operating with losses. A similar behaviour is observed for banks in a study by Ortega and Hommes, who estimate that returns of the banking system fell between 1979 and 1984.[5]

The link between industrial enterprises and banks affected the flexibility of the economic system. This was observed from 1980 to 1983 in Colombia, as well as in Argentina and Chile. Traditionally, enterprises offset losses caused by price declines by lowering dividends, and banks provide loans only to activities that have the conditions to repay them. These mechanisms of adjustment weakened to the extent that enterprises became dependent on bank credit, and started incurring new debts for the amortization and interest payments of past loans. In this way, the bad situation of enterprises, instead of decreasing the demand for credit and reducing the interest rate, actually increased indebtedness.

This framework has introduced great instability into the economic system. On the one hand, the dependence on short-run credit affected the solvency of enterprises, interfering with their long-run plans and damaging their financial position. In Colombia, the increase in indebtedness coincided with the growth of substandard loans in bank portfolios, reducing their capacity to assume risks. Bankruptcies increased and the government had to intervene, giving generous lines of credit in order to avoid the liquidation of large enterprises. On the other hand, the concentration of bank portfolio was contrary to the principle of risk diversification and made economies more vulnerable. A small failure in the system was enough to cause the fall of the group and generate a chain effect on economic activity.

Clearly, financial liberalization did not stimulate competition between a great number of economic agents. In practice, it generated instead monopolistic competition that ended in greater concentration. It was observed that a great number of conglomerates in many countries were able to dominate and control a substantial part of many economies.

Interest Rate

The financial liberalization made resources accessible for a greater number of activities, increasing the demand for credit and raising the interest rate. How-

Table 5. Stock prices and interest rates in Argentina, Colombia, and Chile

Year	Argentina		Colombia		Chile	
	Stock prices	Interest rate	Stock prices	Interest rate	Stock prices	Interest rate
1970						
1971	61.0		80.9		76.1	−15.4
1972			48.3		57.8	−23.5
1973			39.0		52.0	−14.8
1974	180.5	−40.8	37.1		41.9	−12.4
1975	189.7	127.1	17.1	6.8	30.2	−67.6
1976	313.1	17.7	76.7	8.2	30.8	−62.0
1977	576.6	39.1	43.6	−0.4	33.3	15.9
1978	1,141.8	35.1	36.9	7.6	45.6	0.4
1979	1,326.0	16.6	66.5	11.4	50.4	−2.2
1980	2,443.8	12.0	56.8	11.8	22.8	5.7
1981	1,963.2	38.7	27.0	13.4	39.8	19.3
1982	1,486.2	35.1	20.3	13.7	25.1	11.4

Sources: Stock prices: as for table 4. Interest rates: Colombia, Banco de la República; Argentina and Chile, Joseph Ramos, *Estabilización económica en el Cono Sur*, Estudios e Informes de la CEPAL, no. 38.

ever, the levels observed in practice were higher than the ones predicted by models of perfect competition. The interest rate on loans exceeded for several years the estimates of the marginal productivity of capital. Furthermore, the assumption that this variable could be regulated easily through the supply of credit proved unfounded.

It can be observed from tables 5 and 6 that the financial collapse that began between 1979 and 1980 generated a fall in stock and urban property prices that was not accompanied by a fall in interest rates in Colombia, Argentina and Chile. This behaviour can be interpreted easily using the elements developed in the first section. At the end of the speculative period, the asset prices had risen more than they should have and a great deal of liquidity was required to maintain them at those levels. Indebtedness had to be increased to finance the loans of the past plus the interest rate payments. Thus, the resistance to selling assets at lower prices resulted in a demand for credit greater than the one prevailing during the boom. In fact, the tendency of the real interest rate to settle above the marginal productivity of capital, both in expansion periods as well as in contraction periods, is a strong argument in favour of the hypothesis that it is determined by speculative factors. The demand for liquidity is generated by the efforts to obtain profits as well as to prevent losses.

Now it is possible to consider the relation between high interest rates and the strength of economic groups. The origin of both phenomena is similar. On the

Table 6. Real price indexes of urban property in Argentina, Colombia, and Chile

Year	Argentina	Colombia	Chile
1969			
1970			
1971			
1972			19.3
1973			
1974			68.0
1975			
1976			60.9
1977	70.9		
1978	80.9	64.5	
1979	85.1	81.9	
1980	100.0	100.0	100.0
1981	67.4	111.6	221.3
1982		108.6	

Sources: Colombia, Centro National de la Construcción (CENAC); Argentina, Roque Fernández, "La crisis financiera Argentina, 1980–1982" (unpublished); Chile, José Pablo Arellano, *De la liberación a la intervencíon: El mercado de capitales en Chile 1974–1983* (CIEPLAN, Santiago, 1984).

one hand, we saw that groups that owned banks found that the most profitable alternative was to use savings deposits to acquire existing and related enterprises. At the same time, groups that did not own financial intermediaries tried to buy banks. In this process, credit emerges as a central element for developing profitable operations. The demand for credit increased and put pressure on the interest rate. On the other hand, we saw that, while the stock market was dominated by the regular stockholders, prices were highly sensitive to interest rates. However, this relation is diluted to the extent that groups acquired the equity of regular stockholders and took control of the enterprises. The decisions became insensitive to transitory changes and dependent on the long-run returns to monopolistic power. The adjustment is not governed by short-term returns and current prices, but by a large positive elements game. In these conditions, the decrease in the profit of enterprises does not lead to a process of reduced credit demand. For this reason, due to the attitudes of groups with respect to credit, the demand for credit is not very elastic to the current interest rate.

At this point it is easy to summarize the implications of the financial transformation for the interest rate. The high levels cannot be attributed only to the removal of administrative restrictions: they are mainly related to the phenomena generated by the liberalization of a market subject to rigidities and distortions. Individuals move to purchase enterprises that are in a clear disequilibrium situation, in the sense that the quantitative restrictions of liquidity cause their prices to settle below their real value. The greater demand for these enterprises, together with the presence of adaptive expectations in stock markets, generated a price rise that was interpreted as a signal of similar movements in the future. Under these conditions, the interest rate may reach any level and is not very elastic to credit variations. Not even the financial collapses were able to push it to reasonable levels.

Efficiency, Savings and Investments

The financial liberalization was justified on grounds of efficiency.[6] On the one hand, it was considered that a high interest rate would stimulate an increase in savings. On the other hand, it was assumed that free movement of resources and interest rates would induce competition that would allocate resources to the most profitable activities. It was also supposed that the economic group organizations were highly efficient at channelling resources to activities of high risk and great size. Within this framework, a rise in savings, an increase and improvement in investment, and an acceleration of the rate of growth were predicted.

The defenders of financial liberalization had an optimistic conception of the relationship between savings and interest rate. This relationship has been the subject of an inconclusive debate that it is unnecessary to recall in this paper. The difficulties lie, in part, in the fact that changes in the interest rate, being relatively small, did not generate observable effects on savings. Although it was known that a variation of one or two points in the interest rate did not affect consumers' spending habits, the impact of fluctuations of six or seven points was unknown. In fact, the magnitude of the rise in interest rates that occurred during the financial liberalization provided a laboratory setting to elucidate this question.

In Sarmiento, it is shown that the influence of the interest rate on the savings rate in Colombia is small.[7] The fluctuations in savings come mainly from the external sector and fiscal policy.

The figures suggest a similar behaviour for Argentina and Chile. Table 7 indicates that the saving rate did not increase in the three countries between 1975 and 1982, when the interest rate reached the highest values.

These results contrast with the evolution of financial instruments. One observes in table 8 that these instruments increase significantly in Colombia. The same behaviour is observed in Chile and Argentina.

How can the behaviour of financial instruments and global savings be reconciled? The first answer is that individuals shifted from one type of savings to another. In fact, in Colombia, bank financial instruments substituted for stocks, direct investments in limited corporations, and financial instruments of the illegal market. Furthermore, the expansion of financial savings came, in a high proportion, from the income generated by high interest rates. This is equivalent to a transfer from enterprises to savers. Finally, enterprises paid the interest rates through a reduction in profits.

Table 7. Rates of national savings and interest rates, Argentina, Colombia and Chile

	Argentina		Colombia		Chile	
Year	Savings	Interest rate	Savings	Interest rate	Savings	Interest rate
1970	20.4		17.8		15.7	
1971	21.5	−15.4	16.6		13.1	
1972	21.6	−23.5	16.9		6.5	
1973	21.4	−14.8	18.2		7.9	
1974	19.8	−12.4	19.5		16.8	−40.8
1975	18.5	−67.6	16.6	6.8	7.6	127.1
1976	22.8	−62.0	18.1	8.2	11.7	17.7
1977	26.2	15.4	20.1	−0.4	8.4	34.1
1978	23.9	0.4	19.5	7.6	8.1	35.1
1979	21.2	−2.2	19.0	11.9	11.6	16.6
1980	18.7	5.7	18.8	11.8	14.1	12.0
1981	15.0	19.3	17.0	13.4	7.8	38.7
1982	15.4	11.4	17.6	13.7	0.5	35.1
1983	12.8			19.0	5.9	

Sources: Savings: ECLAC. Interest rates: Argentina and Chile, Joseph Ramos (see table 5); Colombia, Banco de la República.

Table 8. Ratio of financial instruments to Gross Domestic Product, Colombia[a]

Year	Annual balances (GDP)	Annual increments (GDP)
1971	6.5	0.7
1972	6.4	0.4
1973	8.7	2.6
1974	8.8	0.7
1975	9.7	1.5
1976	13.0	3.6
1977	19.3	2.0
1978	15.2	1.7
1980	19.1	4.8
1981	21.9	4.8
1982	22.3	0.0

a. Financial instruments include all savings deposits and bonds; they do not include stocks.
Source: Banco de la República and estimates of the author.

These adjustments affected the behaviour and relations of economic agents. The decision to buy a stock or a share in a partnership implied a certain type of investment decision in the past. Naturally, the financial liberalization, having stimulated the shifts towards short-term titles, favoured savings channelled by financial institutions. This transformation weakened the link between savings and investments. Resources that previously went directly to investments came to depend on the discretion of financial intermediaries who, in many cases, found it profitable to divert them to other activities. Thus, financial liberalization increased the credit used to acquire durable consumer goods, to support speculative activities, and to purchase existing enterprises.

None of this is to reject the possibility that the rise in interest rates contributed to an increase in the savings of some economic agents. But this increase, if it occurred, was offset by the shift of credit towards consumer durable goods and speculative activities. Really, the main effect of financial reforms was not to increase savings, but to widen financial intermediation. The great mistake lies precisely in assuming that the two elements are equivalent.

The freedom to channel savings did not mean that they were directed to the most efficient activities. The resources moved towards the activities that were previously restricted. Moreover, the greater access to credit did not generate more competition among economic agents, but greater flexibility and power for economic groups. These groups used their banks to orient credit to acquire enterprises whose price was depressed by liquidity restrictions and enterprises that increased their monopolistic power. Finally, the preference for risk of the economic groups did not show up in the development of large investments, but rather in a clear preference for speculative activities and transfers.

3. Conclusions and Policy Implications

The theory developed in the first section allows one to identify the special characteristics of the economic system that differ significantly from the classical assumptions about capital markets. Some of these peculiarities were observed clearly during the financial liberalization process and also later during the financial crisis. In this concluding section, the information that supports empirically the presence of these elements and its general implications on the functioning of the economic system are summarized.

The complementarity between physical assets and credit weakened many of the simplifications of the orthodox vision. The tendency to discuss the supply of and demand for funds and the interest rate independently from the rest of the economic system cannot be justified when physical assets and the financial markets are closely related.

The relationship between liquidity and large enterprises is confirmed by factors of different kinds. First, the access of individuals to credit made the existing enterprises more tradeable. The increase in indebtedness that took place between 1976 and 1980 coincided with an increase in stock transactions and the stock price rise.

The direct relationship between the price of stocks and the interest rate observed during a long period is also evidence of the complementarity. This behaviour would not be observed under equilibrium conditions. It is the result of a disequilibrium condition due to complementarity, low elasticity of supply of existing enterprises and some durable consumer goods, and the prevalence of adaptative expectations.

The presence of powerful economic groups denies support to the classical assumption that capital markets are the result of competition between a great many individuals. The action of groups within a large positive elements game generates monopolistic competition and an increased demand for credit. Furthermore, the traditional role of the banking system to regulate and select this demand is weakened by a framework of related enterprises and financial intermediaries. The demand for credit becomes excessive and the financial market and resources tend to concentrate, increasing the economy's vulnerability.

The operation of groups within a large positive elements game was corroborated during the financial crisis. The bankruptcies of banks and enterprises were not covered by profits generated in the past or in other activities. The government had to intervene to capitalize the banking system to offset portfolio losses and to concede subsidized credit. On the other hand, small savers incurred losses owing to the inability of financial institutions to pay back their deposits or to the fall in stock prices. There is also some evidence that this game leads to a situation in which many do not survive. In many countries a small number of economic groups ended up dominating the economy.

The general assumption that the financial market is stable and adjusts smoothly did not hold. The evolution of stock prices and interest rates from 1976 to 1982 in Colombia, Chile, and Argentina suggests the contrary. The system can operate under disequilibrium for long periods because of the complementarity between credit and physical assets and the inelasticity of supply of large enterprises and some durable consumer goods. Furthermore, the system may be highly unstable because of the presence of monopolies and the prevalence of adaptative expectations.

The hypothesis that financial liberalization contributes to increases in savings and investment rates was not validated. The figures of Colombia, Argentina, and Chile do not show any relation between savings and interest rates from 1973 to 1982.

Also, it was not true that savings move automatically towards more efficient investments. Capital markets have characteristics that stimulate them to move towards existing enterprises and durable consumer goods.

The simultaneous presence of these ingredients has created a financial structure different from the one that results from traditional conceptions. The complementarity between large enterprises and credit, the operation of groups within the context of a large positive elements game, and the predominance of adaptative expectations result in a highly unstable financial market without the capacity to generate signals that lead resources to the most efficient activities. The system is exposed to bank portfolio concentration and to a process of asset price inflation and interest rise that can lead the economy to collapse. Private returns are not generated in efficient activities, but rather by exploiting imperfections and speculative opportunities. On the other hand, the inelasticity of savings to interest rate, the fact that the elasticity of demand for durable goods to the interest rate is lower than the elasticity of demand for entrepreneurial investment, the low elasticity of supply of existing large enterprises and the monopolistic power of groups do not permit resource movements to productive investment. The liberalization of funds and the interest rate does not guarantee the best resource allocation.

Under this framework government intervention is required to assure the stability of financial markets and to move resources towards the most efficient activities.

Clearly, the failures of financial markets are due to the characteristics of the economies and to institutional factors. The former are inherent in the system and they are associated with complementarities, low elasticities, and slow market adjustments. In general, little can be done to modify them. The most that can be done in practice is to moderate the negative effects. In contrast, institutional factors that are particularly related to financial groups could be altered and corrected.

The disequilibrium situation generated by the complementarity between liquidity and financial assets and their low elasticity cannot be corrected through an increase in the money supply. The distortion can be removed only by government intervention in the allocation of resources, applying the traditional instruments of selective credit. Development funds should be oriented to channelling resources towards new investments. Furthermore, bank borrowing should be subject to strict regulation to avoid its being diverted towards existing properties and speculation.

Government intervention should be extended to cover interest rates. There is no reason for maintaining high interest rates: they do not stimulate savings but, on the contrary, restrict investment and raise production costs. Furthermore, the efficiency of the market mechanism is limited in controlling variables such as the interest rate, which is greatly influenced by economic distortions and speculative conditions. The effective control of interest rate requires the application of administrative mechanisms. However, the experience of the decade of the 1960s shows that negative interest rates generate excess credit demands that impede monetary control. Under these conditions, it seems that real interest rates should be positive, perhaps, but no more than one or two points.

The conditions that grant privileges and monopolistic possibilities to certain groups should be corrected. The privilege of managing a public good should be regulated. Activities that require state intervention as a last resort should be defined clearly and submitted to regulations that allow government action when it is proved that these activities are managed incorrectly.

The formation and organization of groups should also be regulated. It is convenient to establish mechanisms and conditions that compel them to respond jointly to the losses in the different activities that they administer and control. Thus, the large positive elements game will become a more even game.

It should be recognized that the ownership link between the financial and the industrial sector results in activities that are highly effective for speculation. The removal of the link would not affect the capacity of the economy to develop great investment projects; on the contrary, it would stimulate self-sufficient corporations to undertake productive activities. It would be a decisive step towards the establishment of a stable financial market.

The operation of financial and industrial activities separately would require a market of longer-maturity financial instruments. On the one hand, some kind of government intervention is necessary to induce financial institutions to issue long-term bonds; on the other, the stock market should be developed. The dividend policy should be regulated to guarantee that it reflects the real situation of enterprises. Tax measures that discriminate against stocks should be removed and, perhaps, replaced by others that stimulate them. Finally, the institutional conditions for developing secondary markets should be created.

The usefulness of associations within the financing sectors is not so clear. Experience shows that these associations tend to acquire some institutions with the loans of others, generating a concentrated and vulnerable financial system. However, the solution is not to promote a structure based on independent institutions. The general tendency is to provide banking services together, and there is some evidence that it generates great economies of scale. The practical alternative is, rather, to develop rules that define the degree of collusion among institutions and forbid the use of credit to acquire the assets of the same group. In this way, the instability factors would be removed, preserving the advantages of economies of scale.[8]

The links between the industrial sector are determined by natural factors. The dynamic of industrial development generates learning and skills that can be applied in related investments. Interference with this natural process retards the expansion and diffusion of technology. Even so, the practice of acquiring related activities cannot be left completely to the discretion of the private sector. As in the financial sector, these associations should be clear and should imply some kind of solidarity.

Appendix. Expected Return of Stocks

The return of a stock is equal to the dividend plus the increase of its price in real terms. This valorization is unknown and introduces uncertainty to the market. The elements used for individuals to infer it requires the application of some elementary mathematics.

The expected return of a stock is equal to the dividend plus the expected valorization.

$$\frac{D}{Pa} + \left(\frac{\dot{P}a}{Pa} - \frac{\dot{P}}{P}\right)$$

where D = dividend, Pa = stock price, P = general price index, $\frac{\dot{P}a}{Pa}$ = rate of change of stock prices, and $\frac{\dot{P}}{P}$ = rate of change of general price index.

Suppose that the interest rate on bonds adjusted by risk is r. In equilibrium we will have the following equation:

$$\frac{D}{Pa} + \frac{\dot{P}a}{Pa} - \frac{\dot{P}}{P} = r$$

If the term on the left were less than that on the right, individuals would shift from stock to bonds, generating a fall in stock prices. In the opposite situation, the stock price would rise. The change in stock price, at the same time, would affect the dividend and the expected valorization. Each variable is the cause and the effect of others.

This process can be simplified by introducing some practical elements. Suppose that the stock market is operating in equilibrium conditions. If the dividend rises less than inflation, the stock return falls. Individuals will attempt to sell them until the point at which the rate of decline of the real stock price falls is equal to the rate of decline of the real value of the dividend. This interpretation of market behaviour is relatively simple and it is known for most of the economic agents that operate in the stock market. In fact, it can be supposed that individuals form their expectations on the basis that the rate of change of dividends is equal to the rate of change of prices. In algebraical terms:

$$\frac{\dot{P}a}{Pa} = \frac{\dot{D}}{D}$$

Thus, in equilibrium we will have:

$$r = \frac{D}{Pa} + \frac{\dot{D}}{D} - \frac{\dot{P}}{P}$$

The right-hand term of this equation can be estimated with the historical data of dividend and inflation. The results of this exercise are summarized in table 2 of the text.

Notes

1. R. McKinnon, *Money and Capital in Economic Development* (Brookings Institution, Washington, D.C., 1973), chap. 6; D. Levhary and Don Patinkin "The Role of Money in Simple Growth Model," *American Economic Review*, September 1968.
2. Eduardo Sarmiento, *Funcionamiento y control de una economía en desequilibrio* (CEREC and Contraloría General de la República, Bogotá, 1984), chap. 6.
3. In all games there are winners and losers: the former reflect the positive elements and the latter the negative ones. In a zero sum game, which is usually the case in sports, the positive elements are equal to the negative elements. The groups play a game of sum zero, or slightly positive, and have high probability of winning in many activities – so they are in a game which offers them significant positive elements.
4. See note 2 above, chap. 7.
5. Francisco Ortega and R. Hommes, *Estado, evolución y capitalización de bancos y corporaciones financieras* (Convención de Asobancaria, Cartagena, 1984).
6. Ronald McKinnon (note 1 above); Edward Shaw, *Financial Deepening in Economic Development* (Oxford University Press, London, 1973).
7. See note 2 above, chap. 8.
8. O. Bernal and S. Herrera, "Producción, costos y economías de escala en el sistema bancario colombiano," *Ensayos sobre política económica*, March 1983.

COMMENTS ON THE MEXICAN FINANCIAL SYSTEM

David Ibarra

1

There is a certain irony in discussing financial liberalization with reference to the Mexican case. In any event, I would like to present some general remarks on past trends and more recent structural changes in Mexico's financial sector.

As a point of departure I would emphasize the fact that over the period 1956–1972 the Mexican economy was successfully integrating itself in the international financial community while achieving a remarkable rate of growth.

The financial system was open in the sense of having free exchange convertibility and absence of any other control over international financial flows. However, with a minor exception, no foreign bank could operate in competition with Mexican intermediaries for domestic cash and time deposits within national boundaries. By the same token, foreigners were excluded – again with some exception – from participating in the equity of banking institutions.

In order to stabilize such an open system, the Mexican authorities maintained a virtually fixed exchange rate with respect to the US dollar, and were therefore obliged to control inflation within a range similar to that of the American economy (see table 1). Equilibrium in the monetary market was – and in a way still is – a function of the Central Bank's ability to cover government deficits, diverting resources from the private to the public sector, without crowding out effects on business investment.[1] Otherwise, growth would come to an end and strong pressures would dangerously upset prices and the balance of payments, with the risk of generating capital flight.

In other words, the openness of the financial system combined with a fixed rate of exchange implied an imperfect control by the Central Bank over the

Table 1. Exchange rate

Year	Average rate (pesos per dollar)
1970	12.50
1975	12.50
1976	19.95
1977	22.74
1978	22.72
1979	22.80
1980	23.26
1981	26.23
1982	148.50
1983	161.35
1984	210.00

Source: Banco de Mexico, *Informes anuales.*

money supply or the rate of interest. Therefore, economic stability was strongly linked to fiscal policy and to the politics behind it.[2] In addition, stability was crucially dependent on the efficiency of the financial system in mobilizing internal savings or in borrowing from abroad.

Much of the credit allocation was made by administrative means rather than through the market. Legal reserve requirements and other devices were used to shift resources for the financing of public deficits. The resulting scarcity of loanable funds, and the need to devote resources to high-priority activities, paved the way for direct credit allocation, either by means of the regulatory instruments of the Central Bank (financial intermediaries integrated their portfolios in accordance with instructions issued by the Bank of Mexico) and the operations of public and development banks, or through ad hoc mechanisms (governmental trust funds, for instance).

The need to dampen real or potential inflationary pressures associated with governmental deficit spending, and, of course, the influence of vested interests in a protective commercial policy, explain a good deal of the persistent bias to overvalue the currency, to undervalue the price of public services (or the sale price of commodities produced by state enterprises), and to grant cost subsidies instead of higher prices to activities subject to price control authorizations.

2

The built-in policy and political proclivities of the Mexican economic model reached a critical point in the mid-1970s. By 1976, external shocks, cumulative peso overvaluations, postponement of tax and banking reforms, agricultural stagnation, and the decreasing growth effects of industrial import substitution broke the delicate financial balance, setting the stage for a deep economic crisis.[3] While the rate of growth diminished on the average from 1970 to 1976, social and political tensions were on the rise. The outcome was no other than compensatory government spending. The public deficit began to rise at an impressive pace: 100 per cent in 1972, 36 per cent in 1973 and 60 per cent in 1975. Since then, the instrumental association, as seen by the political establishment, between government legitimacy, growth, and public spending, has become clearly visible.

Pressured by mounting and conflicting demands on the part of well-organized groups, authorities used the budget to give them satisfaction, even at the cost of self-defeating indebtedness and instability in the longer run. The overwhelming fact has been the need to strengthen a political legitimacy more and more wedded to the possibility of attaining, on the one hand, not only continuous but fast economic growth, and, on the other, economic justice. On that account it is worth noting that after the 1930s the dominant ideology shifted – without passing away entirely – from social reform to growth politics.

The 1976 crisis was finally overcome through the adjustment of the exchange rate, the reduction of the public deficit as percentage of GDP (from 1977 to 1980 it fell below the 4.8 per cent level of 1976: see table 2), and tight monetary policy. Of course, great help came from the sharp increase in oil exports and prices. However, government expenditures kept up a fast rate of growth, hand in hand with private investment. Therefore, some of the oil income had to pay for the increased budget requirements, as well as for the resulting inflated demand for imports.

3

However, starting in 1978, for the first time in many years, the government attempted to deal with the financial constraints of the economy through various structural or institutional reforms.

On the fiscal side a comprehensive tax reform was implemented in several stages, running from 1977 to 1981. With the enactment of a value added tax law and a system for the co-ordination of federal, state, and municipal taxes, it was possible both to increase fiscal revenues and to simplify greatly an old-fashioned tax structure. More than 600 different local taxes and more than 30 federal taxes were consolidated, and a unified tax system was established all over the country for the first time since Independence. Instead of overlapping fiscal revenue laws, a system of participation in common income tax fund was used to channel fiscal resources to each of the three levels of government.

In December 1980, the income tax reform was approved by Congress. Its

Table 2. Federal government revenue, spending, and deficit (percentage of GDP)

Year	Revenue	Spending	Deficit
1970	9.6	11.6	2.0
1973	10.2	13.4	3.3
1974	10.5	13.7	3.2
1975	12.3	16.5	4.2
1976	12.6	17.4	4.8
1977	13.2	16.7	3.5
1978	13.8	16.8	3.6
1979	14.3	17.4	3.1
1980	17.0	20.0	3.0
1981	17.1	23.8	6.7
1982	17.9	30.4	12.5
1983	18.5	—	—

Sources: Secretaría de Hacienda, *Estadisticas de finanzas publicas* (Mexico City, 1981) and Banco de Mexico, *Informes anuales* (1982).

Table 3. Consolidated income and expenditure of the public sector (billions of pesos)

	1981	1982	1983	1984[a]
Income	2,045	3,617	7,769	12,885
Expenditure	2,841	5,146	9,195	14,779
Other expenses	69	129	95	141
Deficit	865	1,656	1,521	2,035

a. Estimates.
Source: ECLAC.

main objectives focused on the globalization or aggregation of incomes for tax purposes, the elimination of tax privileges related to specific activities, and the introduction of adjustments to correct some of the inflationary distortions on taxable income (adjustment of income brackets, depreciation allowances, capital gains, etc.). To complete the fiscal legislation overhaul, a new fiscal code was enacted, as well as a Federal Duties Law and New Customs Law.[4]

As a joint result of the fiscal reform and of expanding oil revenues, federal government income increased from 12 to 18 per cent of GDP between 1976 and 1982, with a significant improvement in fiscal elasticity (see tables 2, 3, and 4).

A banking reform was also badly needed. Note that the internal and external upheavals of the 1970s were producing negative processes of financial intermediation (the sum of current accounts plus saving deposits in banks to GDP decreased from 32.5 to 22.8 per cent between 1972 and 1979: see table 5.[5] On

Table 4. Federal government income (billions of pesos)

Year	Total revenues	Income tax	Sales tax	Other taxes	Oil taxes
1977	231	94	39	79	18
1978	304	134	48	95	27
1979	413	173	73	120	45
1980	684	247	120	153	164
1981	935	339	158	203	234
1982	1,532	465	217	393	457
1983	3,181	737	543	741	1,170
1984[a]	4,381	1,029	827	977	1,548

a. Estimates.
Sources: Secretaría de Hacienda and Banco de Mexico.

Table 5.

Year	Financial intermediation[a]
1970	30.0
1971	31.7
1972	32.5
1973	29.2
1974	26.4
1975	27.3
1976	21.0
1977	20.2
1978	22.4
1979	22.8
1980	23.0
1981	23.9

a. Current accounts plus saving deposits in private banks.
Source: Banco de Mexico, *Informes anuales*.

the other hand, the financial intermediaries' institutional organization was far from modern and not conducive to growth. The small size of many banks and financial institutions and their excessive specialization, either by type of operation (commercial credit, mortgages, savings and loans, etc.) or by sector of economic activity, precluded the supply of adequate and efficient financial services to the economy. Governmental banks had a method of funding their operations that was heavily dependent upon the fiscal resources of foreign borrowing, for they lacked a network of branches inside the country through which they could mobilize domestic savings. Lastly, the stock exchange and the capital national markets were underdeveloped given the increasing demand of the Mexican economy for financial services.

In response to these shortcomings, steps were taken towards the establishment of full service or integrated banks. The government actively promoted mergers and consolidations. The number of private banking institutions was reduced from more than 400 to less than 50 in the period 1978 to 1982. Within the governmental banking sector, a process of mergers and accommodations was set in motion as of 1977. Two of the national development banks took over two corresponding commercial banks, and the remaining public institutions were directed to develop or strengthen their network of branches on a nationwide scale. As noted, the intended purpose was to reduce the dependency on foreign resources for funding domestic operations, but also present was the important objective of stimulating competition in the financial markets by providing new or alternative banking services all over the country.

Analogous efforts were devoted to strengthening the stock exchange market.[6] In 1977, treasury certificates were issued for the first time; six years later, their outstanding value exceeded 500 billion pesos. The treasury certificates were followed by issues of petro-bonds, commercial paper, banking acceptances, and convertible bonds, among other things, that greatly improved the range of financial instruments and options offered by the stock market. In addition, the Mexican government set up a Central Deposits Institute, gave fiscal incentives to transactions operated through the stock exchange and took the initiative of modernizing laws and regulations applicable to that market.

However, either the financial reforms did not have the time to produce all their expected effects or they proved insufficient to cope with the combination of overambitious economic goals and external shocks; the plain fact was that a new wave of instability was set in motion from the beginning of 1982 (see tables 6, 7, and 8).

Attempting to speed up growth, increase employment, and improve income distribution and welfare, Mexico spent its oil revenue and resorted to additional external borrowing. From 1978 to 1981, GDP increased at more than 8 per cent (table 9), employment grew by 5 per cent a year, and heavy investments were directed to build up oil and petrochemical production facilities (roughly 30 per cent of the public capital budget).

The fast rate of demand expansion and the overvaluation of the peso pushed imports to record levels (the import coefficient almost doubled between 1978 and 1981). The same factors induced the Mexican business community to borrow heavily from international capital markets. During the bonanza, the domestic financial system had obvious limitations in coping with the fast-expanding demand for loanable funds, and the currency overvaluation reduced the cost of external financing, while the oil revenue was taken as a safeguard against the risk of devaluation. In that respect, it is worth noting that the fast

Table 6. Consumer Price Index (1978 = 100)

Year	Percentage change (December–December)[a]
1977	20.7
1978	16.2
1979	20.2
1980	29.8
1981	28.7
1982	98.2
1983	80.8
1984	59.2

a. The data between the periods 1977–1980 and 1981–1984 is not strictly comparable, owing to modifications in the price index.
Source: Banco de Mexico, *Informes anuales*.

Table 7. Balance of payments (millions of US dollars)

Year	Current account	Trade balance	Factor income
1976	−3,005	−2,572	−1,800
1978	−2,693	−1,227	−2,162
1979	−4,870	−2,188	−3,194
1980	−7,223	−2,057	−4,703
1981	−12,544	−3,110	−7,331
1982	−5,922	+5,584	−11,599
1983	+4,968	+13,952	−9,095
1984[a]	+3,818	+13,808	−10,125

a. Preliminary estimates.
Source: Banco de Mexico, *Informes anuales*.

Table 8. Consolidated public deficit (percentage of GDP)

Year	Deficit
1977	6.7
1978	6.7
1979	7.4
1980	8.0
1981	14.7
1982	17.6
1983	8.6
1984[a]	6.9

a. Estimates.
Source: ECLAC.

Table 9. Gross Domestic Product (1970 prices)

Year	GDP (billions of pesos)	Rate of growth
1975	609.9	
1976	635.8	4.2
1977	657.7	3.4
1978	711.9	8.2
1979	777.2	9.2
1980	841.9	8.3
1981	908.8	7.9
1982	903.8	−0.5
1983	861.8	−4.7

Source: Banco de Mexico, *Informes anuales.*

increase in the government's current expenditure (between 60 and 70 per cent per year from 1977 to 1981) did not generate crowding-out effects on the private business sector because of the generous supply of external resources (in 1981, foreign borrowing reached a peak of 24 billion dollars: see tables 10 and 11).[7]

4

For a while, the country did attain its development goals, but at the price of creating domestic financial imbalances and greater external vulnerability.[8] From mid-1981, the world's recession had a significant impact on the Mexican economy. Oil prices declined as well as income from other export products, and interest rates were extremely high by historical standards. Equally significant was a dramatic shift in expectations from the optimism of the bonanza to a grim outlook concerning the feasibility of sustaining growth, reducing inflation, and exercising control over the exchange rate.

In February 1982, capital flight and pressures on the exchange market could not be resisted. The currency was allowed to float and by March it had moved from 26.4 to 45.9 pesos per dollar (see table 1). Inflation intensified as result of a huge increase in salaries authorized by the government, higher prices of imports, and distorted expectations. Capital flight, reduction of the supply of foreign loans, and internal restrictions created an acute scarcity of funds available for credit financing.

Public expenditures were cut and rates of interest increased, but in spite of these and other governmental measures the economy could not regain its equilibrium. In August 1982, persistent instability in the exchange market led the authorities to establish a two-tier exchange rate. During the same month, the dollar-denominated deposits in Mexican banks (Mex-dollars) were made

Table 10

	Pesos/percentage growth									
	1978		1979		1980		1981		1982	1983
Total liabilities	1,211	31	1,590	36	2,158	54	3,317	126	7,541 127	12,111 61
Mexican pesos	751	33	1,000	40	1,396	46	2,042	99	4,076 100	6,796 67
Foreign currency	460	28	590	29	762	67	1,275	169	3,465 172	5,315 53
Total financing	1,145	26	1,442	36	1,956	53	2,991	127	6,796 127	10,341 52
To non-banking private sector	495	67	829	4	866	38	1,197	30	1,568 31	2,389 52
To non-banking public sector	650	−6	613	78	1,090	65	1,794	192	5,228 191	7,952 52
Money supply (M1)	260	33	346	33	461	34	616	44	885 44	1,310 48

Source: Banco de Mexico, *Informes Anuales*.

Table 11. External debt (billions of dollars)

Year	Total	Public debt	Bank debt	Private debt
1976	26	20	1	5
1977	30	23	2	5
1978	34	27	2	5
1979	40	30	2	8
1980	51	34	5	12
1981	75	53	7	15
1982	81	59	8	14

Source: Secretaría de Hacienda and Banco de Mexico.

payable only in local currency. And in September, the two most drastic – at least by Mexican standards – economic measures of the post-war period were taken: the creation of exchange controls and the nationalization of private commercial banks.[9]

Those decisions were a major departure from the traditionally liberal policy on financial affairs followed by Mexican governments. Issues of an economic and political nature were at stake. Even if the measures taken might prove to be technically wrong, the authorities felt compelled to put an end to the capital outflows and to recuperate the degree of freedom in economic policy decisions lost through the slow historical process of integration with world financial markets. Politically, the government also felt the need to state publicly, even loudly, its determination to stop exchange speculation because of its devastating effect on the population's standard of living.

And yet, nationalization of the banks had side-effects that created new social cleavages. Those effects were related to policies that went beyond the limits of the agreed social consensus between the state and significant segments of the polity. Moreover, in the Mexican case, private bankers, as a group, used to play a strategic role as political intermediaries between the government and the business community. Through formal and informal mechanisms, the Mexican Association of Bankers actively participated in the discussion of macroeconomic policies, and its members somehow managed to influence and to co-ordinate the different groups of interest within the private sector.

5

As a concluding remark, let me say that economic polity history after 1982 is, perhaps, the account of government efforts to fully regain business confidence, while adjusting the economy to a harsh internal and external environment.[10] On the financial side it is clear enough that any attempt to liberalize regulations

Table 12. Public debt service as a percentage of current account revenues (millions of US dollars)

Year	Debt service	Current revenues	%
1977	3,847	9,177	42
1978	6,287	11,926	53
1979	10,174	16,403	62
1980	7,681	25,021	31
1981	10,282	30,809	33
1982	11,558	30,717	38

Source: Banco de Mexico.

will have to wait for the consolidation of the nationalized banking system and perhaps for more permanent solutions to the international debt problem (see table 12). Of course it is hard to anticipate the future. External political influences and internal demands from the social groups that are burdened with the adjustment costs might possibly change the present policy trends.

Notes

1. L. Solis, *Economic and Policy Reform in Mexico* (Pergamon Press, New York, 1981).
2. P. García Alba and J. Sierra, "Causas y efectos de la crisis económica de México" *Jornadas*, 104 (1984).
3. W.R. Easterly, " Fiscal and Financial Effects of Devaluation: A Macro Model of Mexico," mimeo (Trade and Development Workshop, MIT, Cambridge, Mass., 1984).
4. Secretaría de Hacienda, *El sistema impositivo mexicano* (Mexico City, 1981); F. Gil-Díaz, "Lessons from Mexico Tax Reform," mimeo (Mexico City, 1985); T.J. Kehoe and J. Serra-Puche, "A Computational General Equilibrium Model with Endogenous Unemployment," *Journal of Public Economics*, 22 (1983); A. Ortiz-Salinas, "La política económica de la política fiscal," paper presented to the UN Workshop on Fiscal Policy, Kyoto, Japan, 1985.
5. Banco de México, *Informes anuales*, various issues; and CEPAL, *Estudio económico anual*, various issues.
6. R.D. Johnston, "Should the Mexican Government Promote the Country Stock Exchange?" *Inter-American Economic Affairs*, 26 (1972).
7. P. García Alba (note 2 above); Joseph Kraft, *The Mexican Rescue* (Group of Thirty, New York, 1984).
8. S. Kalifa Assad, "Inversión privada y política gubernamental en México," paper presented to the Conference on Investment Processes in Mexico and the USA, Stanford University, Calif., January 1985.
9. C. Bazdresch, "La nacionalización bancaria," mimeo (Mexico City, 1984).

10. A. Calderón and A. Buenrostro, "Algunas consideraciones sobre el programa inmediato de reordenación económica," Document No. 3, UNAM, Programa Universitario Justo Sierra (Mexico City, 1984); World Bank, *Mexico, Recent Economic Developments and Prospects*, Report No. 4996-ME (World Bank, Washington, D.C., 1984).

A NOTE ON OUTWARD-LOOKING POLICY AND GROWTH PERFORMANCE: THE REPUBLIC OF KOREA AND SELECTED LATIN AMERICAN COUNTRIES

Hirohisa Kohama

Development Policies: Open or Protective?

The differences in the basic orientations of national development strategies and their bearing on actual growth performance constitute one of the important issues in development, and have often been the subject of theoretical and empirical discussions. The question focuses on whether outward-looking liberalization is more effective in ensuring long-term economic development than inward-looking protection, or vice versa. More specifically, import substitution and export promotion are often contrasted as two basic policy alternatives for industrialization and economic growth.

Most of the studies undertaken so far seem to conclude that open and liberalized development policies have been more successful in generating favourable economic performance on a sustained basis than closed, protective ones. However true this generalization might be, a particular policy mix desirable for a given country should be formulated by taking note of the country's typological characteristics: namely, natural resource endowments, sizes of population and national economy, access to the markets of industrialized nations, historical backgrounds and institutional capacities, and so forth.

The appropriate policy mix should vary, moreover, in accordance with the state of development of a given country. Different stages of development naturally require different strategies to sustain growth on a long-term basis. If one considers the temporal, or historical, dimension of economic development, the difference between import substitution and export promotion in industrialization strategy might be viewed as one of continuity or sequence rather than of contrast or opposition. The key question would then be when and how to switch from one policy mix to the other with as few repercussions as possible.

Apropos of trade and industrialization policies, many "latecomer" countries, including Japan, began their initial period of modern economic growth by exporting primary commodities to purchase manufactured consumer goods from industrialized countries. The process would then typically shift to domestic manufacturing of consumer goods, or primary import substitution, to be followed sooner or later by the start-up of primary export substitution. The next step is secondary import substitution, or domestic production of intermediate and capital goods, which would then shift to secondary export substitution. This typical process of modern economic development has been observed in such resource-poor nations as Japan and the Republic of Korea, but in many other countries the shifts may overlap rather than neatly dovetailing in time.

In contrast to the experiences in Japan and the Asian NICs like the Republic of Korea, the courses of economic development observed in Latin American countries did not conform, in many cases, to the typical pattern mentioned above. Advanced developing countries in Latin America such as Argentina, Brazil, Mexico, and Chile launched industrialization much earlier than the Asian NICs like Korea and Taiwan. In Argentina, for instance, the manufacturing sector already contributed over 23 per cent to GDP as early as 1950, when the share of manufacturing was less than 25 per cent in Japan. Even allowing for the devastating effects of the Second World War on the Japanese economy, there is no denying that Argentina was then well advanced in its industrialization drive. The share of manufacturing in GDP was also relatively high, around 20 per cent, in Brazil, Mexico, and Chile in the beginning of the 1950s.

The import dependence on consumer goods (the ratio of imports to total domestic demand) in four Latin American countries declined to less than 10 per cent in the early post-war years. In Mexico, for example, the ratio of exports to total domestic output had been higher than the ratio of import dependence by the latter half of the 1950s, indicating that the country had completed the process of primary import substitution by then.

What is at issue is this: What became of this successful early import substitution? As already mentioned, industrialization was launched in the Asian NICs much later than in the four Latin American countries, but the pace of subsequent industrialization was notably rapid in the former. Even though the four Latin American countries completed their respective primary import substitution much earlier, their processes of import substitution lasted noticeably longer than in Asian NICs. Such extended periods of import substitution have been closely related to the continued importance of primary commodities exports in these countries; namely, grains and beef in Argentina, coffee, soybeans, sugar, and iron ores in Brazil, copper in Chile and cotton and oil in Mexico.

With regard to development strategies for developing nations, the World Bank appears to favour, on the whole, outward-looking over inward-looking policy mixes and considers the former as more efficient in generating growth and more flexible in absorbing, and adjusting to, external shocks. The Bank's country economic reports often contain recommendations to that effect, and its *World Development Report 1984* explicitly emphasizes the importance of the outward-looking policy mixes for liberalization.

Presumably, the Bank's basic standpoint is that export substitution policies serve to eliminate market distortions created by the protective import substitution strategy, and thereby bring about more efficient resource allocation and restructuring of production. Discontinuation of protective policies forces the domestic industries of developing countries to face international competition, but the resultant shake-out of non-viable firms and the survival of competitive ones serves to transform the domestic structure of production to a more efficient and dynamic one in the long run.

The aim of this paper is to probe the validity and implications of this standpoint. Recent attempts at liberalization in Chile and some other countries have apparently fallen widely short of such expectations, and this calls for some reassessment of the position. At the present time, however, it is beyond the author's scope to discuss exhaustively the significance of the differences in basic development strategy *vis-à-vis* actual economic performances. The present note limits itself to the suggestion of a working hypothesis on the relationship between resource endowments and the effectiveness of alternative development strategies, by comparing the experiences of the Republic of Korea, a notable East Asian NIC, and of four Latin American countries.

Growth Performances

Table 1 compares the macro-economic performances of the five countries under study. Per capita income in 1982 was estimated to be slightly less than US$2,000 in Korea, whereas those in the four Latin American countries were well over US$2,000, the highest, Argentina, amounting to a little more than US$2,500. Brazil and Mexico far outdistanced Korea in GDP for the same year, reflecting their large population bases. In other words, the domestic markets of the two countries were overwhelmingly larger than Korea's. However, the GDPs of Argentina and Chile were smaller than that of Korea. In terms of annual economic growth rates, Korea decidedly outpaced the Latin American countries: growth performances in Argentina and Chile especially were markedly poor during the 1970s.

Table 2 compares the progress in industrialization by the ratio of manufacturing value added in GDP. Rapidity of industrialization in Korea is all too obvious from the table. The ratio of manufacturing value added (Ym/Y) in

Table 1. Macro-economic performance (US$ million, US$, %)

	GDP, 1982	Population (millions), 1982	GDP/N, 1982[a,b]	G (GDP)[a] 1960–1970	G (GDP)[a] 1970–1982	G (GDP/N), 1960–1982[a]
Republic of Korea	68,420	39.3	1,910	8.6	8.6	6.6
Argentina	64,450	28.4	2,520	4.3	1.5	1.6
Brazil	248,470	126.8	2,240	5.4	7.6[c]	4.8
Chile	24,140	11.5	2,210	4.4	1.9	0.6
Mexico	171,270	73.1	2,270	7.6	6.4	3.7

a. G (X) is growth rate of X.
b. GDP/N is per capita GDP.
c. 1970–1980.
Source: World Bank, World Development Report 1984.

Table 2. Industrialization ratio (percentages)

	1955	1960	1965	1970	1975	1980	1981	1982
Republic of Korea	6.1	12.1	17.2	21.7	26.0	28.8	29.1	28.6
Argentina	21.7	23.3	25.8	27.0	27.8	24.8	22.2	22.3
Brazil	21.6	24.9	24.7	27.4	28.3	28.8	27.4	27.1
Chile	21.7	23.2	25.8	26.0	21.5	22.2	21.7	19.1
Mexico	18.0	18.4	20.7	22.9	23.5	24.1	23.8	23.3

Sources: ESCAP, Statistical Yearbook for Asia and the Pacific, various issues; Bank of Korea, Monthly Statistical Bulletin, December 1984; ECLAC, Statistical Yearbook for Latin America, 1983.

Korea remained lower than in the Latin American countries up to 1970. In 1955, it was a mere 6.1 per cent compared with over 20 per cent in the latter countries. By 1975, however, Korea's ratio became more or less equivalent to the levels in Latin American countries, and even surpassed the latter in the early 1980s. It is especially notable that the ratios show a declining tendency in Argentina and Chile during the early 1980s. This reversal of the positions was no doubt because of the wide disparities in the growth rate of manufacturing. According to World Bank estimates, the growth of the manufacturing sector in Korea recorded 17.6 per cent per annum in real terms during the 1960s, compared with 5.6 per cent in Argentina, 5.5 per cent in Chile and 10.1 per cent in Mexico during the same period. For the period of 1970–1980, the annual growth rate was 14.5 per cent in Korea, compared with −0.2 per cent in Argentina, 7.8 per cent in Brazil, −0.4 per cent in Chile, and 6.8 per cent in Mexico.

Changing Structures of Trade

This section compares the growth of manufactured trade and international competitiveness in five countries under study, and then examines changing patterns of export competitiveness in manufactured goods as classified by end use.

Growth of Manufactured Trade

Table 3 shows the trend of manufactured trade by the ratios of manufactured exports (and manufactured imports) to total exports (and total imports) in each country *vis-à-vis* the world.

Before starting a comparison of the ratios of manufactured exports, some explanation is in order. Manufactures are defined here to comprise SITC categories from 5 to 8, but excluding 68 (non-ferrous metals). In other words, the ratio of manufactured exports in Chile, for example, gives a substantially lower share than usual. Moreover, because of the limitations of UN trade statistics, the most recent year in which data is available varies by country; namely, 1981 for Korea and Brazil, 1980 for Argentina, 1979 for Mexico and 1978 for Chile. With regard to exports to the world as a whole, the ratio of manufactures in the most recent year was as high as 90.0 per cent in Korea, distantly followed by 39.1 per cent in Brazil, 23.1 per cent in Argentina, 21.5 per cent in Mexico, and 10.1 per cent in Chile. As already seen from table 2, the percentage contribution of manufacturing to GDP was more or less equivalent in Korea and four Latin American countries. However, the ratio of manufactures to total exports was about 40 per cent in Brazil, and a little over 20 per cent in Argentina. This significant difference between Korea and the four Latin American countries partly reflects the variance of natural resource endowments between these countries. It is not easy to assess total natural resource endowments in quantitative terms for cross-country comparison. But if one supposes that the ratio of manufactures in total exports as weighed relative to the percentage of manufacturing in GDP reflects differences in resource endowments, it is not difficult to understand that the adoption of similar development strategies might result in widely varying growth performances.

When the ratios of manufactures to total imports in the most recent year are compared, we see that they were higher than the same ratios in total exports in Argentina, Mexico, and Chile; that is, except for Korea and Brazil, the countries under study were more or less net importers of manufactures. To examine this point at some length, net export ratio for commodity i as a proxy of *ex post* international competitiveness is calculated:

$$C_i = (X_i - M_i)/(X_i + M_i)$$

The value of C_i varies from -1 to $+1$, with positive values indicating that a given country is a net exporter, and negative ones that it is a net importer. Rises

Table 3. Manufactured trade ratio (percentages)

	Export					Import			
Year	Republic of Korea	Argentina	Brazil	Chile	Mexico	Republic of Korea	Argentina	Brazil	Chile	Mexico
1962	18.8	3.2	3.1	3.5	15.5	56.9	80.1	57.4	66.1	83.2
1963	45.2	5.7	3.0	3.9	17.0	50.6	78.3	55.8	63.2	80.1
1964	46.6	7.3	5.3	3.6	15.9	49.7	67.7	47.8	64.1	82.5
1965	59.3	5.6	7.7	3.9	15.0	51.6	62.3	50.3	63.7	82.4
1966	60.6	6.2	7.1	4.6	18.6	58.5	64.7	53.1	66.4	82.8
1967	66.6	8.1	9.8	3.7	18.8	61.4	68.7	55.3	65.8	83.2
1968	73.9	12.1	8.1	3.1	20.3	63.3	69.6	59.7	68.3	85.6
1969	76.2	13.6	9.7	3.8	25.5	57.3	69.9	64.1	68.3	83.8
1970	76.7	13.9	13.2	4.3	32.5	54.8	73.1	68.4	64.0	80.6
1971	81.8	15.1	15.2	5.2	38.1	54.2	72.1	68.9	64.8	81.6
1972	83.7	20.3	18.8	5.3	35.1	57.2	74.6	71.0	57.1	80.8
1973	84.0	22.4	19.6	3.7	41.9	55.6	67.7	65.4	59.1	73.8
1974	84.6	24.4	24.2	4.4	37.8	51.7	64.0	59.9	45.9	67.6
1975	81.4	24.4	25.3	10.0	31.1	50.7	67.4	62.0	56.3	74.0
1976	87.6	24.8	23.0	10.4	29.1	52.8	65.4	54.3	55.3	79.5
1977	84.8	23.9	25.1	10.7	28.1	52.4	68.8	52.1	62.9	76.6
1978	88.3	26.1	33.3	10.1	27.0	58.5	72.8	50.7	58.9	77.5
1979	88.9	24.1	37.6	—	21.5	55.2	67.0	44.4	—	76.8
1980	89.6	23.1	37.2	—	—	43.1	77.3	40.8	—	—
1981	90.0	—	39.1	—	—	43.3	—	36.2	—	—

a. Manufactured goods: SITC 5–8-68. The manufactured trade ratio is a ratio of export (or import) to total merchandise export (or import).
Sources: IDE, AIDXT.

Fig. 1. Net export ratio of manufactured goods (after IDE, AIDXT)

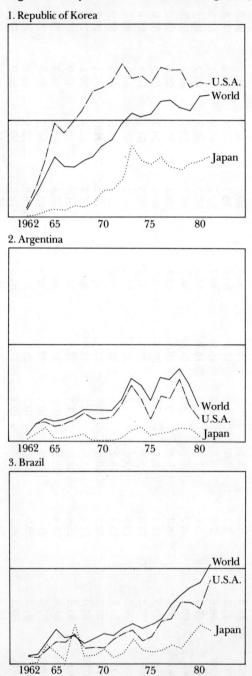

1. Republic of Korea

2. Argentina

3. Brazil

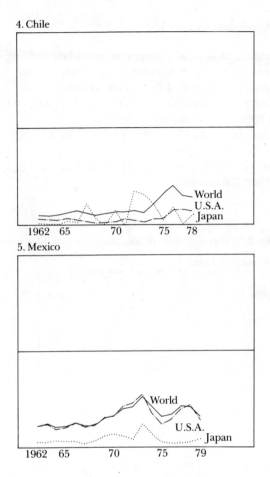

in the value of Ci can be taken to imply improvements in external competitiveness of a given country, as shown in figure 1. The value of Ci for Korea turned positive in 1968 in relation to the United States, and in 1973 *vis-à-vis* the world. Concerning its exports to Japan, the value of Ci still remained negative, but it clearly shows a rising trend.

As for the four Latin American countries, the values of Ci were invariably negative, the only exception being Brazil, in which the value of Ci *vis-à-vis* the world turned positive in 1981. Moreover, while Brazil shows a steady improvement in competitiveness in manufactured exports in relation to the world, the United States, Japan, and the three other Latin American countries do not show clear trends. In the case of Argentina, its competitiveness seemingly improved until 1978, but deteriorated sharply afterwards, thus making it difficult to draw a firm conclusion. In Chile and Mexico, the values of Ci were very low, as well as showing no definite trend of improvement.

External Competitiveness by End Use

In order to examine structural changes in manufactured exports, the same indexes of competitiveness are calculated separately by end use in figure 2. Manufactured goods are classified by end use as follows.
- *Consumer non-durables:* SITC 553, 571, 654, 656–657, 831, 841–842, 851, 863, 892, 895, and 899. (13 commodity categories).
- *Consumer durables:* 666, 696–697, 724–725, 732–733, 812, 821, 864, 891, 893–894, and 896–897 (15 categories).
- *Labour-intensive intermediate goods:* 611–613, 631–633, 651–653, 655, 662–667, 691–694, and 698 (20 categories).

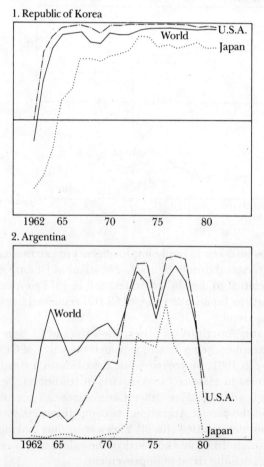

Fig. 2. Net export ratio of consumer non-durables (after IDE, AIDXT)

1. Republic of Korea
2. Argentina

OUTWARD-LOOKING POLICY AND GROWTH PERFORMANCE   131

3. Brazil

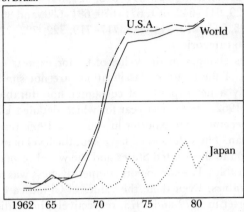

4. Chile

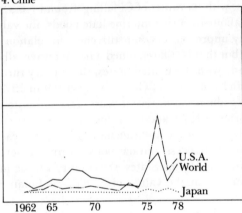

5. Mexico

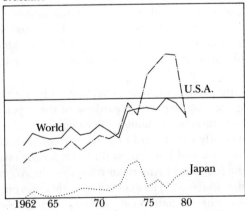

- *Capital-intensive intermediate goods:* 512–515, 521, 531–533, 541, 551, 554, 561, 581, 599, 621, 629, 641–642, 661, 671–679, 681–689, and 862 (38 categories).
- *Capital goods:* 695, 711–712, 714–715, 717–719, 722–723, 726, 729, 731, 734–735, and 861 (16 categories).

Figure 2 compares changes in the value of C_i for exports of consumer non-durables (the rest of the end-use classifications are not similarly illustrated). Korea was already a net exporter of consumer non-durables to the United States as early as 1962, the earliest year for which detailed UN trade statistics are available. It became a net exporter in 1963 and 1965, respectively, *vis-à-vis* the world and Japan. The values of C_i have kept the level of over 0.9 since 1965 and 1973 in relation to the United States and the world respectively.

In consumer durables, Korea's export competitiveness was not as strong as in consumer non-durables. Even then, the values of C_i indicate that the country turned a net exporter in 1972 and 1973, respectively, in relation to the United States and the world. *Vis-à-vis* Japan, Korea still remained a net importer of consumer durables, with its values of C_i falling around -0.5.

In the trade of labour-intensive intermediate goods, the values of C_i indicate that Korea steadily improved its competitiveness in relation to the world and the United States, but that is still remained a net importer, albeit small, relative to Japan. In its trade with the United States, the country turned a net exporter as early as 1964, and the value of C_i rose to over 0.9 in 1972, although it declined afterwards to about 0.5 in recent years. In relation to the world, Korea became a net exporter in 1970, and the values of C_i have risen steadily since.

Concerning capital-intensive intermediate goods, Korea shows improvements in export competitiveness, but has not yet turned a net exporter. In relation to the world and the United States, the country's trade in these goods was more or less balanced, but the values of C_i *vis-à-vis* Japan remained around -0.5 in recent years.

Production of capital goods requires the most advanced technologies among the five groups of manufactures classified by end use. Consequently, Korea's export competitiveness in capital goods remained negative throughout the period under observation in relation to the world, the United States, and Japan. But it must be noted that the country's values of C_i clearly show a rising trend. The calculated values in recent years were around -0.2 in relation to the world and United States, and -0.75 *vis-à-vis* Japan.

From the foregoing discussion, it is important to note that Korea steadily improved its export competitiveness, regardless of the types of manufactured goods and of their destination, although the calculated values of C_i varied in absolute terms. Especially with regard to consumer non-durables, the country's C_i rose very close to $+1.0$ and has kept at this level for the last few years.

Let us turn to the four Latin American countries. In Argentina, as already mentioned in relation to figure 1, its export competitiveness in consumer goods and labour-intensive intermediate goods improved up to the mid-1970s and

OUTWARD-LOOKING POLICY AND GROWTH PERFORMANCE   133

then plummetted sharply afterwards. With respect to capital-intensive intermediate and capital goods, there is no perceptible trend, and the values of Ci did not rise. Concerning its trade with the world and the United States, the country once became a substantial net exporter of consumer non-durables and labour-intensive intermediate goods, but regressed to net importer in later years. This sort of setback is not very common in other newly industrializing countries.

Among the four Latin American countries under study, Brazil has been most dynamic in its improvement of export competitiveness in manufactured trade. Except for its trade with Japan, the values of Ci definitely show a rising trend. Although the country remained a net importer of capital-intensive intermediate and capital goods, it turned a net exporter of consumer non-durables in 1971 and of consumer durables in 1972 in relation to the world and the United States. Apropos of labour-intensive intermediate goods, it became a net exporter in 1972 *vis-à-vis* the world and in 1976 in relation to the United States.

Because the UN trade statistics for Chile are available only up to 1978, it is not possible to show the effects on international trade of the reported serious deterioration of the country's economy in recent years. Excluding capital-intensive intermediate goods, Chile's values of Ci remained significantly low, on the one hand, and they do not show any appreciably rising trends, with the exception of labour-intensive intermediate goods, on the other. It must be remembered that the indexes of competitiveness by end use include non-ferrous metals (SITC 68) contrary to the calculated values for figure 1, and that Chile's capital-intensive intermediate goods include its sizeable exports of copper and other non-ferrous metals. It has been reported that the index of manufacturing production in Chile dropped from 100 in 1980 to less than 70 in the first quarter of 1983. Drastic liberalization of the economy since the late 1970s must have had something to do with this deterioration. When the statistical information is available for the period after 1978, the country's competitiveness in manufactured exports might show a pattern very similar to Argentina's.

The trend in Mexico seems somewhat different from the other Latin American countries under study. The values of Ci for labour-intensive intermediate goods were already zero at the beginning of the 1960s in relation to the world and the United States, while those of capital-intensive intermediate goods *vis-à-vis* the world and the United States show a somewhat declining trend of competitiveness.

It is not very common to observe a decline in competitiveness in developing countries which have reached the stage of being called NICs. Mexico is considered as one of the leading Latin American NICs, on a par with Brazil. As already mentioned, Brazil shows a clearly rising tendency of export competitiveness, although its values of Ci were appreciably lower than Korea's. Supposing that the basic attributes of the newly industrializing countries derive from their dynamic improvements in competitiveness of manufactured exports, the foregoing findings about Mexico imply, in the author's view, that the country falls short of being fully qualified as an NIC.

Concluding Remarks: Typological Differences and International Competition

Provided that institutional build-ups are adequate to sustain a given country's manufacturing capability in competitiveness, efficient management, technological innovations, and other requirements, open and liberalized policies would be a better alternative for generating long-term economic development than closed and protected ones.

Post-war Japan initially adopted protective policies in its external economic relations and gradually liberalized them apace with economic recovery and growth acceleration. Japanese domestic industries had time to develop and sharpen their international competitiveness under the protective policy framework, because the domestic market itself was always highly competitive. The protective industrial administration by MITI was often alluded to as the reason for the rapid growth of heavy and chemical industries in Japan during the 1950s and 1960s. And many overlook the important fact that severe competition in the domestic market functioned to accelerate manufacturing enterprises' capability for generating technological innovations. This oversight is a serious mistake, especially when Japan's experience is being discussed for the significance it has for the developing countries.

It is hardly possible to say that the opening up of domestic markets is always good or bad of itself. If the liberalization of a given economy is judged likely to drive domestic industries into bankruptcy, the opening up must be considered to be too early and be tempered accordingly.

The competitiveness of manufacturing industry is central to the choice of outward- or inward-looking development strategy. In countries like Japan and Korea, which are poor in natural resources, it is not possible to continue financing industrialization by exports of primary commodities. The development and continued expansion of heavy and chemical industries must be supported by foreign exchange earnings from exports of light-industry goods. In resource-rich countries, in contrast, exports of primary commodities can pay for the sizeable imports of capital and intermediate goods necessary for industrialization. Such countries have a definitely better head start, but in the long-term their rich endowments seem to inhibit a sustained pursuit of efficiency in manufacturing. In this sense, resource-poor countries might be better placed in their commitment to the continued strengthening of competitiveness.

The Structure of Capital Markets in Asia

ECONOMIC DEVELOPMENT AND FINANCIAL LIBERALIZATION IN THE REPUBLIC OF KOREA: POLICY REFORMS AND FUTURE PROSPECTS

Kim Joong-Woong

Introduction

The rapid expansion of the Korean economy, for the last two and a half decades, has brought about major structural changes in the pattern of industrial activity and external transactions. Since the Republic of Korea launched its First Five-Year Economic Development Plan in 1962, its economy has grown at an average annual rate of 8.2 per cent. As a result of this rapid growth, Korea's per capita GNP in 1984 reached US$2,000, and today it is recognized as one of Asia's leading newly industrializing countries.

Economic growth has not always been smooth. Towards the end of the 1970s, the Korean economy began to show signs of strain from excess demand, and the combination of relatively high inflation and rising unit labour costs resulted in the deterioration of export competitiveness. This situation was greatly compounded by a lethargic world economy in the wake of the second oil crisis in 1979 and culminated in the unprecedented negative economic growth of 1981, which was accompanied by high inflation and a large increase in the balance of payments deficit.

In response to the prospect of slower growth and rising protectionism abroad, the government in 1980 started to implement wide-ranging adjustment programmes, including the adoption of flexible exchange rates, reduction of the budget deficit, and a decreasing monetary growth rate. These policies were initiated in an effort to arrest the inflation rate and strengthen the external position of the economy. The government believed that such adjustment policies were a prerequisite for sustained growth, though they might retard the early recovery of the economy.

Table 1. Macro-economic development and policy response

	1970–1974	1975	1976	1977	1978	1979	1980	1981	1982	1983	1984
Macro-economic indicators											
GNP growth rate (%)	8.8	6.9	14.1	12.7	9.7	6.5	−5.2	6.2	5.6	9.5	7.6
Inflation rate (%) (GNP deflator)	17.7	25.7	20.7	15.7	21.9	21.2	25.6	15.9	7.1	2.9	4.0
Gross investment as % of GNP	26.7	30.0	25.6	−27.7	31.2	35.6	31.3	29.1	27.0	27.8	30.0
National savings as % of GNP	17.9	19.1	23.7	27.5	28.5	28.1	21.9	21.7	22.4	24.8	27.4
Current account balance (mil. US$)	−834.5	−1,886.9	−313.6	12.3	−1,085.2	−4,151.1	−5,320.7	−4,646.0	−2,649.0	−1,606.0	−1,362.1
As % of GNP	6.7	9.0	1.1	0.03	2.1	6.7	8.7	6.9	3.7	2.1	1.7
Unit labour cost (1980 = 100)[a]	18.7	31.4	42.2	54.9	65.7	83.6	100.0	112.2	128.2	134.8	131.4
Import price index (1980 = 100)[b]	45.9	57.1	58.8	59.3	61.9	78.4	100.0	104.0	98.7	94.7	94.4
Export price index (1980 = 100)	53.6	59.3	66.9	71.2	81.3	95.6	100.0	103.0	99.2	96.4	98.1
Rate of increase of exports (%)	48.7	10.8	56.2	28.6	26.5	15.7	17.1	20.1	1.0	11.1	13.5
Policy response indicators											
Nominal exchange rate	370.1	484.0	484.0	484.0	484.0	484.0	607.4	681.0	731.1	775.8	807.1
Real effective exchange rate (1980 = 100)	81.8	82.2	88.1	90.9	96.3	103.9	100.0	103.1	99.2	93.9	90.5
M_1 rate of change (%)	30.7	25.0	30.6	40.7	24.9	20.7	16.3	4.6	45.6	17.0	0.5
M_2 rate of change (%)	28.5	28.2	33.5	39.7	35.0	24.6	26.9	25.0	27.0	15.2	7.7
M_3 rate of change (%)[c]	—	28.4	35.6	42.1	35.8	31.0	33.0	30.8	33.2	21.6	20.0

DC₁ rate of change (%)ᵈ	34.8	32.2	21.7	23.6	45.9	35.6	41.9	31.2	25.0	15.7	13.2
DC₂ rate of change (%)ᵉ	—	47.3	39.5	49.6	46.4	46.6	58.0	41.2	38.7	25.7	32.8
Interest rate (bank lending rate)	21.1	15.5	16.1	16.2	16.9	18.5	22.9	19.2	12.0	10.0	10.0–10.5
Unified budget surplus or deficit (bil. won)	—	−466	−398	−476	−616	−440	−1,174	−2,111	−2,222	−951	−923
As % of GNP	—	4.6	2.9	2.6	2.5	1.4	3.2	4.6	4.3	1.6	1.4

a. Nominal wages in manufacturing divided by manufacturing value-added.
b. Real effective exchange rate = effective nominal exchange rate/relative price.
  Effective nominal exchange rate = $100 \cdot \pi \, (FX_{Kor}^{wi} / FX_i) = 100 \cdot FX_{Kor} / \pi FX_i^{wi}$
  Relative price = $100 \cdot \pi (WPI_{Kor}^{wi} / WPI_i) = 100 \cdot WPI_{Kor} / \pi WPI_i^{wi}$
  FX_i: the indices of i-th country currency-dollar exchange rate.
  WPI_i: WPI for country i; wi: the trade weight for country i; the weights are given by the four countries' trade shares.
  The respective weights are: US = 0.45696; Japan = 0.42809; Federal Republic of Germany = 0.07275; UK = 0.04220.
c. M₂ plus deposits at non-bank financial institutions.
d. DC₁ is domestic credit supplied by monetary institutions.
e. DC₁ plus domestic credit extended by non-bank financial institutions.
Sources: BOK, *Economic Statistical Yearbook*, various issues, and Korea Development Institute, *Quarterly Economic Outlook*.

The financial industry and the government's monetary policy have also been liberalized and reformed. In the earlier stages of development, the use of direct and selective instruments of financial policy ensured an efficient allocation of financial resources. However, as the economy grew in scale and complexity of structure, such policy instruments became increasingly inefficient.

Thus, the basic orientation of the government in recent years has been to allow the market mechanism to function more freely. The primary objective of these liberalization efforts, particularly in the financial sector, has been to promote private initiative and competition through the market mechanism. The purpose of this paper is to review the relationship between economic development and the financial industry, to evaluate the recent financial liberalization in Korea, and finally to examine the prospects for further financial liberalization.

Economic Development and the Role of Finance

The Interrelationship between Economic Development and the Financial Sector

Over the last two decades, the financial sector in Korea has made a significant contribution to the country's economic development by mobilizing domestic savings and allocating investment resources. The growth of the financial sector, however, has not kept pace with the growth in other sectors of the economy.

In noting the patterns of growth in the financial and real sector, some questions naturally arise, i.e. what relationship exists between economic development and the financial sector, and why has the financial sector in Korea lagged so far behind the real sector.

In recent years, the role of financial markets in economic development has been the object of increasing concern from development economists. The early studies viewed financial markets as playing a more or less passive role in economic development. Patrick's view was that "demand-following" financial development through the creation of modern financial institutions and diversification of financial assets and services, as a consequence of developments in the real economy may be too passive.[1] This theory has led to the suggestion that the lack of financial innovation stands in the way of the process of economic growth.

Alternatively, policy-makers should move towards a "supply-leading" phenomenon, and deliberately create financial institutions, instruments, and services in advance of demand, first, to achieve the twin objectives of transferring resources from traditional low-growth sectors to modern and high productivity areas, and, second, to stimulate and promote entrepreneurial development in such sectors.[2]

In short, recent studies since the pioneering work of Goldsmith suggest that the process of financial liberalization and encouragement of efficient markets through "deepening" and elimination of "fragmentation" of markets could improve the process of mobilization of savings, as well as the efficiency of investments. This would also improve resource allocation and raise the return on capital and level of output.

Yet, particularly during the latter half of the 1970s, the Korean government had intervened deeply in financial resource allocation in order to develop what are called the strategic industries, such as the heavy and chemical industry, through the regulation of financial institutions by credit rationing and ceilings on interest rates. This financial policy has led to what McKinnon calls the "repression" of financial development and the fragmentation of markets. This not only leads to considerable resource distortion, but also entails the retardation of the financial sector.

At the initial stage of economic development, of course, direct government intervention could be effective and efficient in resource allocation for investment.

However, as the Korean economy has become increasingly complex in its structure and grown in size, it has become more difficult to manage by non-market mechanisms.

Basically, an economy under the free enterprise system can be developed through vigorous investment stemming from the innovation of private entrepreneurs. The investment is financed primarily by bank loans at the early stage of economic development in a developing country like Korea, since other channels of finance are not often available. As the economy grows and develops, financial innovations will also follow, new financial institutions will be established, and thus the channels of finance will be diversified.

Economic development, therefore, brings about development in the financial as well as the real sector. These two sectors are often compared to the two wheels of a bicycle. They develop side by side and one cannot go far if the other is left behind, regardless of whether one follows the "demand-following" or the "supply-leading" hypothesis.

The pattern of development in Korea seems to have been an exception to this general rule, in the sense that the financial sector appears to have been greatly lagging behind the real sector. The financial interrelation ratio, i.e. the ratio of total financial assets to GNP, is far lower than that of advanced countries and even lower than that of Taiwan, a country at a similar stage of economic development. The total amount of assets of the seven major commercial banks in Korea can only be compared to that of a medium-sized rural bank in Japan. There is good evidence for this point.[3]

The Role of Finance in Economic Development

In order to understand further the causes of the retardation of the financial sector in Korea, we need to review the role of finance in economic development, in connection with the development theories mentioned above.

Table 2. Trend of increase in deposit banks' total assets (billion won, percentages)

	1977	1978	1979	1980	1981	1982	1983	1984	Average rate of increase
Foreign bank branches	541[a] (5.4)[b]	820 (5.7)	1,165 (6.2)	2,417 (9.2)	2,879 (8.4)	3,057 (8.3)	4,210 (8.5)	5,084 (8.6)	(37.7)
	(51.6)[c]	(142.1)	(107.5)	(10.1)	(21.8)	(20.0)	(20.8)		
Commercial banks	4,988 (40.9)	7,061 (49.3)	9,298 (49.1)	11,398 (45.2)	11,753 (46.1)	19,463 (46.2)	21,944 (44.4)	26,378 (44.3)	(26.9)
		(41.6)	(31.6)	(28.4)	(32.0)	(23.5)	(12.7)	(20.8)	
Local banks	885 (8.7)	1,177 (8.2)	1,491 (7.9)	1,883 (7.1)	2,456 (7.2)	3,087 (7.3)	3,862 (7.8)	4,831 (8.1)	(27.4)
		(33.0)	(26.7)	(26.3)	(30.4)	(25.7)	(25.1)	(25.1)	
Specialized banks	3,759 (37.0)	5,284 (36.8)	6,978 (36.8)	10,160 (38.5)	13,091 (38.3)	16,010 (38.0)	19,376 (39.3)	23,197 (39.0)	(27.4)
		(40.6)	(32.1)	(45.6)	(28.8)	(22.3)	(21.0)	(25.1)	
Total	10,174 (100.0)	14,345 (100.0)	18,933 (100.0)	26,398 (100.0)	34,179 (100.0)	42,072 (100.0)	49,394 (100.0)	59,491 (100.0)	(28.7)
		(41.0)	(32.0)	(39.4)	(29.5)	(23.1)	(17.4)	(20.4)	

a. Excludes customer's liability on acceptances and guarantees.
b. Constitutional ratio.
c. Ratio of increase compared with the previous year.
Source: BOK, Monthly Statistics.

Table 3. Financial interrelation ratio[a]

	Republic of Korea					Japan	Taiwan	US
	1965	1973	1976	1979	1983	1982	1982	1982
Financial assets balance/GNP in current prices	0.62	1.30	1.23	1.40	2.16	4.56	3.12	3.71
Primary securities/ GNP in current prices	0.38	0.75	0.73	0.84	1.30	2.32	1.72	2.40
Indirect securities/ GNP in current prices	0.24	0.55	0.50	0.56	0.86	2.25	1.40	1.31

a. Foreign financial assets and trade credit excluded.
Source: Bank of Korea.

The role of finance in economic development, in general, can be divided into two parts. The first goal is to provide an adequate supply of money. The second goal is to perform the function of financial intermediation. The first has something to do with the responsibility of the banks, which is also closely related to the government's monetary policy. Banks provide funds for investment in the early stages of economic development through the creation of credit, because the entrepreneurs do not in principle possess all the funds needed for investment.

Therefore, throughout the course of economic development, the private enterprises incur debts and the banks provide them with investment funds. However, it should be noted that the commercial banks are private business firms with public responsibilities, and thus they have to pursue two conflicting aims – to make profits, and to meet the public demand.

In Korea, banks are placed under the control of the government, which seems to be ignorant of the original nature of banks as private firms, whose business is to make a profit. In other words, the Korean government has intervened extensively in determining the financial resource allocation for investment from its own macro-economic view, not from the banks' or from private enterprise's view. Thus, there can be no innovation in bank management, and no renewed development of the financial market.

The second role of finance regarding economic development is that of financial intermediation. The banks mobilize private savings and relay them to investment. This is considered a non-inflationary method of providing investment funds.

As mentioned above, the financial sector in Korea is considered to be very retarded, if one compares its growth with that of the real sector. What are the

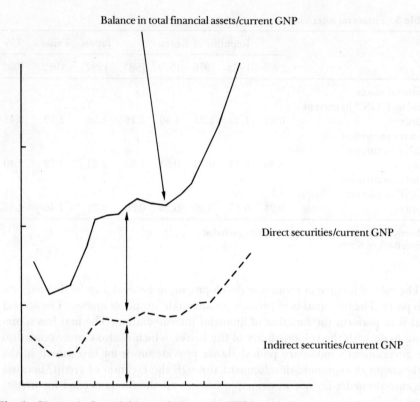

Fig. 1. Change in financial interrelation ratio (FIR), excluding external financial assets and intercompany credit.

causes of this phenomenon? There seem to be several factors that could account for this retarded growth.

The first factor is persistent inflation during the course of economic development. Under the influence of inflation, financial institutions cannot be expected to make significant progress, since the price mechanism cannot work well enough to allocate financial resources efficiently. Furthermore, the banks have not been able to carry out their inherent function of evaluating and reviewing the efficiency and soundness of an investment project. Hyperinflation in Korea may have been caused by inefficient investments as well as by the quantity of money.

The second is lack of innovation in the financial sector due to government control. The banks, owned and operated by the government, have not made any innovations and, as a consequence, a significant amount of loan funds flow outside the organized financial channels.

Third, a great number of investments, especially the government-sponsored ones, have not been efficient enough to produce profits to repay the banks. In addition, many major investments in Korea, particularly during the 1970s, have been financed directly by foreign loans, which has meant a less important role for domestic financial institutions.

As the problems of Korea's financial industry accumulated, it became increasingly apparent that the Korean economy might lose its momentum for continued growth. In response, since 1980 the government has taken reform measures to promote the autonomous management of financial institutions, including the denationalization of commercial banks and the promotion of greater competition in the financial industry.

Evolution of the Korean Financial Market

Until the mid-1960s, the intermediary role of commercial banks was of little significance owing to the negligible savings potential of the nation. The primary purpose of financial institutions until then was to channel funds to rehabilitation projects and farmers. Two special banks, the Korea Development Bank (KDB) and the Korea Agriculture Bank (KAB), accounted for over 70 per cent of total bank lending.

With the introduction of an interest rate reform in September 1965, interest rates were raised substantially and the role of financial institutions as mobilizers of savings was finally recognized. The effect of the 1965 interest rate reform on mobilizing additional domestic resources was remarkable, raising the commercial banks' share of total bank loans from 27 per cent in 1964 to 55 per cent within five years.

Another important role played by the Korean banking system was to facilitate the inflow of foreign loans by guaranteeing repayment. Since 1966, the commercial banks have joined the KDB to become active participants in the massive foreign loan guarantee business.

Along with channelling credit to preferred sectors and projects through existing financial institutions, new institutions were organized to engage in specialized activities that were desired by the government. The number of specialized banks also grew rapidly in the 1960s. The Medium Industry Bank (MIB) and the Citizens' National Bank (CNB) were both created shortly after the military government came into power in 1961. In addition, the Korea Exchange Bank (KEB) and the Korea Housing Bank (KHB) were established in 1967.

However, the interest rate reform of 1965 was too short-lived to sustain the momentum it initially generated. Between 1968 and 1972, bank interest rates were lowered in several stages to reach levels well below the pre-reform rates. After 1970, there were some signs of financial disintermediation through the unorganized money market (UMM) until the Presidential Emergency Decree of August 1972 froze the market.[4]

Efforts in the early 1970s to reduce the curb loan market by establishing new non-bank financial institutions and to promote the capital market contributed significantly to the diversification of Korea's financial market. In connection with the decree, short-term finance companies and mutual savings companies were established, and credit unions were modernized. Although the growth of these institutions has been constrained by various operational restrictions, including interest rates, they seem to have succeeded in attracting funds which otherwise might have been supplied to the curb market.

Until 1972, the Korean capital market was little more than a secondary market for national bonds issued during and after the Korean War. Only a limited number of equity shares were traded in the market, and no corporate bonds were issued. Since then, however, the capital market has grown very rapidly, thanks to strong promotional measures by the government. Together with measures geared to increasing demand for securities, favourable corporate tax treatment has been given to publicly held firms, and the government has been empowered to order selected companies to go public.

The Korean financial sector has been effective in channelling financial resources to the preferred or strategic sectors designated in the economic development plans. This was achieved mainly through preferential loans extended at low interest rates. This process, however, retarded the development of the Korean banking sector. The banking sector suffered from a lack of competition, inadequate incentives, and serious operational inefficiencies, problems that largely stemmed from extensive government intervention:[5]

First, the major commercial banks were all largely public enterprises, with the government owning a majority of their equity shares. The result was direct government control of virtually every aspect of banking operations.

Second, as a means of allocating investment funds according to national priorities, the government resorted to strict control of interest rates and provided extensive preferential credit at subsidized rates.

This process led to rapid monetary expansion, which brought about chronic inflation. Particularly in the latter half of the 1970s, the ambitious undertakings in heavy and chemical industrial projects, together with a large inflow of foreign capital and the improved balance of payments situation, resulted in accelerated monetary expansion.

Institutional Framework of the Financial System in the Republic of Korea

The financial system in Korea consists of a highly regulated formal sector, composed of deposit money banks and non-bank financial institutions, in addition to an unorganized sector known as the "curb market." This dualistic structure is an important characteristic of the Korean financial system.

Formal Financial Sector

The formal financial sector is highly segmented and financial institutions specialize in particular segments of the market. The demarcation between financial institutions is effected by an extensive regulatory framework. Banking regulations focus on separating various financial functions, maintaining a fairly high degree of specialization, and reducing competition among financial institutions. Substantial barriers to entry and to branching exist, and the introduction of all new services or debt instruments require prior approval.

In general, the International Monetary Fund distinguishes the monetary financial system from non-monetary financial institutions, depending on whether the financial institutions can create money or not. The monetary system consists of the Bank of Korea (the central bank), which creates money as the monetary authority, and deposit money banks. All the commercial banks and some specialized banks whose main revenue source is deposit money are classified as deposit money banks.

The formal financial sector is dominated by the banking system, which accounted for about three-fifths of total domestic liabilities in 1984. The banking system is composed of seven nationwide commercial banks, ten local banks, seven specialized banks, and 50 domestic branches of foreign banks. The nationwide commercial banks have an extensive branch network throughout the country, while local banks' branches are confined to the provinces in which their head offices are located. Each specialized bank was established under its own special law – unlike the commercial banks which were established under the General Banking Act – and has its own specifically defined purpose. Among specialized banks, however, the Korea Exchange Bank, the Small and Medium Industry Bank, the Citizens' National Bank, the Credit Department of National Agricultural Co-operatives Federation and member co-operatives, the Credit Department of Central Federation of Fisheries Co-operatives and member co-operatives, and the Korea Housing Bank are all classified as deposit money banks.

On the other hand, the Korea Development Bank, the Korea Export–Import Bank, and the Korea Long-Term Credit Bank, which are all development institutions, are treated as non-bank financial institutions. Foreign banks do not have branch networks in Korea: most branches are located in Seoul. As a consequence, these banks do not have substantial domestic currency deposits.

The importance of the banking system in the formal financial system has declined steadily since the early 1970s. The ratio of bank deposits to total deposits of the financial system fell from 85 per cent in the early 1970s to 57 per cent in 1984. Within the banking system, the seven nationwide commercial banks held about half of total bank deposits in 1984, while specialized banks held about 40 per cent, and local banks held 10 per cent: foreign bank branches held only 1 per cent of total bank deposits.

The above-mentioned development institutions, savings institutions, investment companies, insurance companies, securities-related companies, and other financial institutions are included among the non-monetary financial institutions.

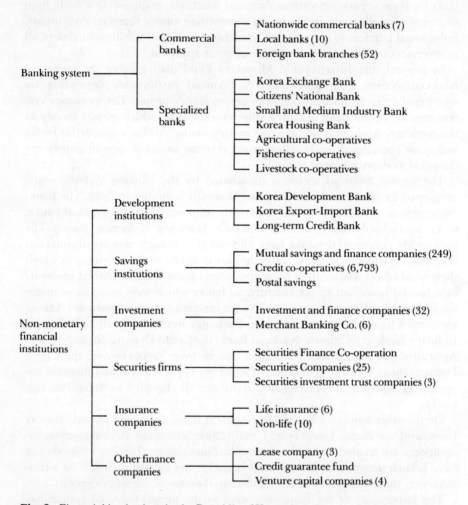

Fig. 2. Financial institutions in the Republic of Korea (numbers in parentheses indicate the number of corresponding banks).

All of the non-monetary financial institutions, except the development institutions, are called non-bank financial institutions (NBFIs).

NBFIs can be classified into four categories according to their principal activities:

1. Savings institutions, including mutual savings and finance companies, trust companies, and credit unions.

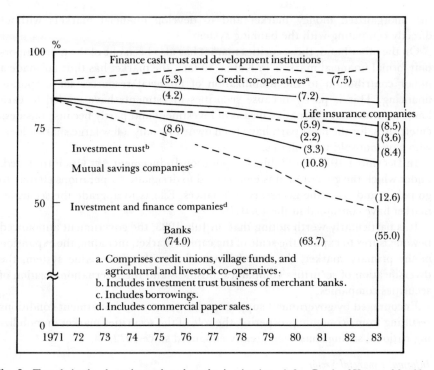

Fig. 3. Trends in the deposit market share by institutions (after Bank of Korea, *Monthly Bulletin*, April 1984).

2. Investment companies, including investment and finance companies, merchant banks, and investment trust companies.
3. Development institutions, including the Korea Development Bank, the Export–Import Bank, the Korea Land Development Corporation, and the Korea Long-Term Credit Bank.
4. Life insurance companies.

NBFIs have been the most dynamic element of the financial system in recent years. Deposits with these institutions have grown by 48 per cent per annum from 1975 to 1984, compared to 28 per cent per annum for the banking system. The rapid growth of NBFIs has been due to the emergence of new financial instruments issued exclusively by these institutions; yields on these instruments are significantly higher than banks' deposit rates.

The fastest-growing segment of the NBFIs has been the savings and investment institutions, particularly mutual savings companies, investment trust companies, and investment and finance companies. These savings and investment institutions were created in the early 1970s to attract funds from the unorganized money market and to develop financial markets without directly competing with the banking system.

On the other hand, the securities market has also been one of the more important formal financial markets in Korea. In fact, the market has thus far made a minor contribution to the mobilization of financial resources for corporate financing. This is partly because securities investments have tended to earn lower returns than alternative means of investment and partly because business enterprises, for the most part, have been owned by only a few large shareholders who have not sold their shares.

In December 1972, the Public Corporation Inducement Act was introduced, under which the government is empowered to designate corporations eligible to go public and issue the necessary ordinances. Efforts to upgrade the securities market have continued in the 1980s.

It is particularly worth noting that, in July 1983, the government announced new measures to expand the scale of the capital market, including the expansion of the primary market, the introduction of a market-price issue system, the diversification of securities companies' business and the internationalization of securities companies.

Encouraged by government support and the improved investment conditions resulting from economic prosperity, the role of the securities market in mobilizing national savings has been greatly enhanced since 1972.

Informal Financial Sector

Loans from the banking system are extended primarily to larger firms for financing fixed capital formation. Consequently, smaller firms, particularly those in need of working capital, are compelled to seek loans in the curb market. This market has several attractions for borrowers, including confidentiality, quick loan approval, and risk assessment based on credit analysis rather than collateral. Large savers, who are attracted by the relatively high rate of return, are the principal suppliers of funds.

The curb market is intertwined closely with the banking system, as funds raised by dealers in the curb market often pass through the banking system. Typically, a potential borrower contacts a broker, who in turn raises funds in the curb market; these funds are deposited in a commercial bank and are then used to finance a bank loan to a borrower designated by the broker. The depositor collects the bank deposit rate plus a premium paid by the broker, while the borrower pays the official interest on the bank loan plus a premium to the broker. Alternatively, the potential borrower could issue commercial paper, or another debt instrument, that would be accepted by a short-term finance company at the official interest rate; this debt instrument would be rediscounted in the curb market with the borrower paying the additional interest rate.

The curb market has lessened the adverse effects of pervasive regulation of the formal financial system, but at a certain cost. The quasi-legal status of the

Table 4. Domestic deposits by financial institutions (billion won)

	1974	1975	1976	1977	1978	1979	1980	1981	1982	1983	1984	Average growth rate (%)
Banking system[a]	2,193 (85.7)[c]	2,792 (84.1)	3,728 (82.1)	5,179 (81.1)	6,883 (79.3)	8,766 (76.1)	11,538 (73.3)	14,916 (68.9)	18,847 (64.6)	22,096 (60.9)	24,817 (57.6)	27.46
Non-bank financial institutions[b]	367 (14.3)	528 (15.9)	813 (17.9)	1,207 (18.9)	1,801 (20.7)	2,746 (23.9)	4,209 (26.7)	6,728 (31.1)	10,331 (35.4)	14,175 (39.1)	18,268 (42.4)	47.81
Investment and finance companies	162	254	438	659	941	1,338	2,098	3,215	4,227	5,497	7,012	47.36
Investment trust corporation	5	15	43	124	241	362	635	1,354	2,768	3,654	4,313	96.60
Mutual savings and finance companies	51	50	68	107	161	268	400	612	957	1,474	1,992	44.27
Life insurance companies	98	128	169	233	351	658	943	1,391	2,209	3,363	4,738	47.38
Others	51	81	95	84	107	120	133	156	170	187	215	15.48
Total	2,560 (100.0)	3,320 (100.0)	4,541 (100.0)	6,381 (100.0)	8,684 (100.0)	11,512 (100.0)	15,747 (100.0)	21,644 (100.0)	29,178 (100.0)	36,271 (100.0)	43,058 (100.0)	

a. Including deposits, money in trust, commercial bills sold, RPs, postal savings, deposits at National Savings Co-operatives, long-term credit time and savings deposits, and accounts of the KDB.
b. Including bills sold and CMA.
c. Numbers in parentheses are composition ratios.
Source: Ministry of Finance, *Public Finance and Banking Statistics* (Seoul, 1985).

curb market has created problems for both lenders and depositors. Enforcement of loan contracts by the courts is difficult, particularly when the interest rate charged exceeds that permitted under usury laws; depositors also face a similar problem. Absence of regulation has made this market susceptible to disruptions stemming from fraud and other illegal activities, such as the incidents that occurred in 1982 and 1983. The repercussions of these scandals in the formal financial sector have placed an added burden on monetary management. Secrecy increases information costs and makes it more difficult for depositors and lenders to evaluate creditworthiness. Finally, fragmentation of the curb market limits its ability to exploit economies of scale. For these reasons, the transaction costs and the risk premium in interest rates are high.

The government has made two major attempts to absorb the curb market into the formal financial system – one in the early 1960s and the other in the early 1970s. In both instances, new NBFIs were created to compete with the curb market. Notwithstanding these efforts, the curb market has continued to be a significant element of the financial system because regulations imposed on these new institutions have limited their ability to compete with it. Nevertheless, NBFIs provide an important link between the banking system and the curb market.

No accurate estimates of the size of the curb market exist. The most reliable ones were those made in 1972, when the authorities attempted to absorb it into the formal financial sector. Data obtained at that time indicated that the market (in terms of its liabilities) represented about one-quarter to one-third of broad money; however, this may understate its actual size because of the desire of many participants to remain anonymous.

Financial Liberalization in the Republic of Korea

Since the early 1980s efforts to enhance the efficiency of the entire economic system have led to important steps toward reform in at least three areas: realignment of the industrial incentive system, promotion of competition between domestic and foreign firms, and financial liberalization.

In restructuring the industrial incentive system, the concept of encouraging "strategic industries" is shifting to support for activities which benefit Korean industry in general, such as technological innovation and manpower development. By being all treated equally, industries are encouraged to compete on their own merits.

Institutional improvements favouring competition include the implementation of the Anti-Monopoly and Fair Trade Act in April 1981 and the liberalization of external transactions to accelerate the drive for industrial efficiency.

To introduce greater competition from abroad, the import liberalization ratio has also been raised from less than 70 per cent five years ago to the current 88

Table 5. Progress of import liberalization, 1975–1984

	CCCN[a]			Import liberalization ratio
Date	Total	Prohibited	Restricted	Automatic approval	

Date	Total	Prohibited	Restricted	Automatic approval	Import liberalization ratio
Dec. 1975	1,097	57	509	531	48.4
Nov. 1976	1,097	54	499	544	49.6
Dec. 1977	1,097	50	456	591	53.9
Dec. 1978	1,097	—	335	675	61.5
July 1979	1,010	—	327	683	67.6
July 1980	1,010	—	317	693	68.6
July 1981	7,465	—	1,886	5,579	74.7
July 1982	7,560	—	1,769	5,791	76.6
July 1983	7,560	—	1,482	6,078	80.4
July 1984	7,915	—	1,203	6,712	84.8
July 1985	7,915	—	970	6,945	87.7

a. 1981–1985, on the basis of CCCN 8-digit.
Source: Ministry of Commerce and Industry.

per cent level. Moreover, application for technology import licences is now automatically approved in most cases. The environment for direct foreign investment in Korea is now much more favourable, with the introduction of the negative list system.

As an essential part of these institutional reforms, there has been a thorough restructuring of the financial sector. This effort has included measures (a) to enhance financial autonomy, (b) to rely on the price mechanism, (c) to promote the universal banking system, (d) to improve financial policies, and (e) to internationalize the Korean financial market.

Greater Autonomy

Greater competition, and therefore greater efficiency, has been promoted by encouraging financial institutions to diversify and by offering them more autonomy in decision-making. The idea is that, ultimately, it will be the private sector rather than the public sector which determines the performance of the financial markets.

One of the characteristics of the Korean banking system prior to the reforms was that commercial banks had been under the control of the government, which was the major shareholder of the four nationwide commercial banks. By March 1983, the government had relinquished its shares of these banks to the private sector. Furthermore, the government revised the General Banking Act in 1982, primarily to provide the denationalized banks with a freer hand in dealing with their own managerial affairs. The other purpose of the revision of the Banking Act was to prevent banks from being placed under the influence of a few large shareholders.[6]

The government is also reducing its overall intervention in bank management. Commercial banks' budgets, personnel management, and organizational structures are no longer subject to approval by the government. Management is becoming more responsible for operational performance. In the management of monetary policy, the government has changed the methods of controlling the money supply to make use of indirect mechanisms, such as the level of the required reserve ratio.[7]

Greater Reliance on Price Mechanism

The government has moved to lower entry barriers to the financial sector by establishing new joint venture banks, short-term finance companies, and mutual savings and loan companies.

Of course, complete management autonomy and efficiency would imply the freedom to set interest rates. However, if the government relinquished control too suddenly, disruptive and destabilizing consequences might ensue. So, as in many other countries, the Central Bank is likely to continue to have strong influence on the general level of interest rates.

Market forces have become a more important factor in the determination of interest rates. The majority of interest rates in Korea have been administered largely by the monetary authorities and the rates thus determined tended to be considerably lower than those prevailing in the private money market. Persistence of high inflationary pressures, particularly in the 1970s, made it almost impossible for monetary authorities to let institutional interest rates be determined by market forces.

Interest rate differentials between general loans and preferential policy loans of banks were eliminated in 1982, as an important step toward de-emphasizing preferential policy loans. Commercial banks have also been freed from the obligation to make preferential policy loans.

Furthermore, in January 1984, a narrow band, from 10 to 10.5 per cent, was introduced in order to permit banks to charge different rates based on borrowers' creditworthiness.[8] At the same time, a new term structure for deposit rates was established to encourage a shift into longer-term deposits.

In November 1984, rates on long-term deposits were raised by one percentage point in order to narrow the gap between institutional and market interest rates. As a first step towards enlarging the scope of free markets, the ceiling on inter-bank call rates was lifted and the rates on issuance of corporate bonds were liberalized, except for those guaranteed by banks.

As an essential element of financial development, interest rate liberalization will gradually be implemented. As inflationary expectations recede with continued price stability, the gap between official and free market rates should continue to narrow. This will allow the government to liberalize interest rates on money market instruments and ease the regulation of bank interest rates. The range of bank lending rates will be broadened, and deposits rates will also be set in a more flexible manner.

Toward a Universal Banking System

The government is in the process of instituting universal banking by integrating segmented financial markets and eliminating restrictions on the scope of banking services. First of all, various new financial instruments were introduced to reorganize short-term and long-term financial markets and to enhance the role of financial institutions as savings mobilizers. Some of the new instruments were shared by different financial institutions to promote competition.

In 1980, sales and purchases of government and corporate bonds on repurchase agreement, which had been handled only on an individual basis, were absorbed into official RP markets with security companies as their dealers. From 1982, the banks were allowed to engage in sales on a repurchase agreement of the bonds they accepted from government and public utilities.

Commercial paper was introduced for investment and finance companies in 1981 and for large securities companies in 1984. Initially, CP rates were allowed to be determined by market forces as a first step towards gradual decontrol of interest rates. However, as the rates thus determined became too high in comparison with administered rates, they became subject to a maximum rate set by the Minister of Finance. Moreover, the Cash Management Account, a Korean version of the money market fund in the United States, was introduced in 1984 for investment and finance companies.

In the banking sector, trust business was opened to all commercial banks to make up for the loss of their market share after the introduction of new instruments for non-bank financial intermediaries. In addition, in 1984 commercial banks were permitted to issue Certificates of Deposit at a fixed discount rate which was higher than that on time deposits with comparable maturities.

In 1981, commercial banks launched retail banking services, including household checking accounts, automatic deposits in monthly salaries, credit cards, and payment of public utility charges through automatic transfers and automatic teller services.

Financial intermediaries have also been allowed to broaden the range of their financial transactions. Large securities companies may participate in some money-market activities, commercial banks may accept mutual savings deposits and engage in factoring, and provincial banks may set up investment trusts.

The distinction between banks and non-bank institutions is therefore becoming blurred, and competition is increasing in financial markets in line with the trend towards a one-stop universal banking system. The diversification of banking business will result in banks being more able to meet the changing demands of customers for financial services.

Improvements in Financial Policies

Monetary and credit management in Korea traditionally relied on credit ceilings and other direct control measures rather than on more orthodox, indirect control tools. The credit ceilings for individual banks led to imbalances in credit allocation among industries and firms, rigidity in credit management, and limited incentives for competition. However, beginning in 1982, an indirect credit control system was adopted in which the total reserve holdings of banking institutions are used as the operational target in controlling monetary growth.

Important progress has also been made toward a more rational interest-rate structure. By June 1982, preferential interest rates applied to various policy loans were abolished to gradually phase out policy loans. Actually, the relative share of preferential policy loans has been declining as a result of government efforts to restrict the growth of the National Investment Fund (NIF).

Finally, in July 1982, the Ministry of Finance announced a proposal requiring all financial transactions to be made in the real names of the persons concerned. This proposal sought to reduce the scope of the curb market and to establish an institutional framework for more equitable taxation. Although this controversial proposal was postponed by the National Assembly until some time after 1986, some progress has already been made toward discouraging people from holding financial assets anonymously or under fictitious names through the imposition of heavier taxes.[9]

Toward Internationalization

Although considerable progress has already been achieved in the internationalization of Korea's banking industry, much remains to be done. In 1981, the government prepared a long-term blueprint to liberalize fully the capital flow in and out of the country by the early 1990s. To this end, the Ministry of Finance announced a schedule for the internationalization of the securities market.[10] Two open-end type investment trusts for foreign investors were also opened in November 1981. It is noteworthy that the Korea Fund, amounting to US$60 million, was successfully set up in May of last year. This fund aims at long-term capital appreciation through investment, by allowing indirect access for leading foreign securities companies to Korea's equity market. Foreign securities firms are now allowed to open representative offices in Korea. Moreover, foreign securities companies will soon be allowed to participate directly in Korea's equity and bond markets.

Table 6. Yearly trend of foreign bank branch openings

	Foreign bank branch										Korean institutions in foreign countries		
	1967–76	1977	1978	1979	1980	1981	1982	1983	1984.3	1985.6	Office total	Branch office	Local corporation
American													
USA	4	3	5		1	3	2			2	(23) 20	(6) 5	(18) 18	(9) 9	(7) 5
Canada			2				1				3	1	0	0	2
Asian													
Japan	4						3	1			(14) 8	(11) 8	(15) 8	(20) 8	(7) 0
India											1	0	0	0	0
Singapore			1			1				−1	3	1	2	4	1
Hong Kong				1			1			−1	2	1	3	6	5
Middle East		1								−2	0	1	2	2	1
European													
UK	1	2	1							1	(15) 6	(1) 0	(11) 6	(7) 3	(0) 0
France	2	1	1							1	6	1	1	1	0
FRG			1						1	−1	2	0	3	3	0
Netherlands				2							1	0	1	0	0
Others													
Total	11	8	11	3	1	4	7	2	2	3	(3) 52	(21)	(2) 46	(19) 55	(2) 16

Sources: Bank of Korea, Ministry of Finance.

In line with the inroads which Korean financial institutions have made overseas, there has been a substantial influx of foreign banks opening branches and offices in Korea. There are 52 foreign bank branches currently operating in the country, the Chase Manhattan Bank being the first to open business in 1967. The purpose of introducing foreign banks into Korea is to facilitate the inducement of foreign capital as well as to motivate domestic banks to improve their banking practices and managerial skills.

National treatment is Korea's policy on foreign banks. In this spirit, the exclusive privileges currently enjoyed by foreign banks will be phased out in step with the removal of restrictions imposed on them. For instance, some discriminatory regulations have already been lifted, and in 1984 foreign banks were permitted to join the National Banking Association for the purpose of exchanging business information with domestic banks. This year, foreign banks have been permitted to enter the trust business and to make use of the rediscount facilities at the Bank of Korea for short-term export financing. In 1986, they will be allowed to make use of the rediscount facilities for all of their operations.

The eventual extension of the open-door policy to the financial sector will permit the entry of foreign banks and non-bank financial institutions on a broader basis in order to help domestic financial institutions upgrade their international banking services.

Future Prospects for Financial Liberalization

Evaluation

When the financial liberalization policy was initiated, there were conflicting opinions on the prospects for success. The sceptics argued that the financial institutions' dependence on government directives and regulations could not easily be ended. In addition, there were many other factors which might prevent financial liberalization. The optimists, on the other hand, believed that since financial liberalization was inevitable anyway, Korea should actively promote it at the earliest possible date. The optimists also recognized that financial liberalization could not be achieved by merely abolishing the regulations and privatizing the financial institutions and that it would occur only when the long-standing dependent relationship between the government authorities and financial institutions was broken.

Although it is still premature to evaluate the performance of financial liberalization, progress over the last five years indicates that the future prospects are good.

In connection with the measures taken so far, there still seem to be some vital problems and constraints that will have to be taken into consideration in order to achieve full financial liberalization.

The problems which the government should pay more attention to are: the independence of the Central Bank; the cost and benefits of financial liberalization; the relationship between the real sector and financial sector; and the liberalization from inside of the financial industry.

The Independence of the Central Bank

As we have previously noted, inflation and financial regulation are vital elements among the various factors retarding the financial market.

From the lessons learned in the 1970s, we are convinced that, in order for the country to continue its high economic growth, we must eliminate inflationary pressures in our economy and achieve financial liberalization. This is one of the reasons why the government launched a comprehensive financial reform.

Table 7. Monetary survey (billion won, percentages)

	1980	1981	1982	1983	1984
Money supply (M2)[a]	12,534.5	15,671.1	19,904.2	22,938.1	24,705.6
	(26.9)[b]	(25.0)	(27.0)	(15.2)	(7.7)
Domestic credit	16,777.6	22,015.7	27,529.0	31,846.7	36,590.1
	(41.9)	(31.2)	(25.0)	(15.7)	(13.0)
Private	16,777.6	22,015.7	27,529.0	31,846.7	36,590.1
	(39.6)	(26.3)	(25.1)	(17.6)	(14.0)
Govt and public agencies	731.1	1,742.3	2,158.4	2,013.1	1,972.8
	(118.3)	(138.3)	(23.9)	(−6.7)	(−2.0)
Net foreign assets (US$million)	−891	−3,295	−5,999	−6,162	−7,291
Quasi-money	8,577.0	11,499.8	14,104.9	16,154.7	17,306.1
	(32.2)	(33.9)	(20.7)	(14.5)	(10.7)
Time and savings deposits	8,577.0	11,499.8	13,659.7	15,672.7	17,306.1
	(31.3)	(34.1)	(18.8)	(14.7)	(10.4)
Money supply (M1)[c]	3,807.0	3,982.4	5,799.3	6,783.4	6,820.7
	(16.3)	(4.6)	(45.6)	(17.0)	(0.5)
Currency in circulation	1,856.4	2,025.4	2,573.7	2,874.4	3,109.4
	(15.7)	(9.1)	(27.1)	(11.7)	(8.2)
Deposit money	1,950.6	1,957.0	3,225.6	3,909.0	3,711.4
	(16.8)	(0.3)	(64.8)	(21.2)	(−5.1)
Total liquidity (M3)[d]	17,791.2	23,277.8	31,004.5	37,706.1	45,236.8
	(33.0)	(30.8)	(33.2)	(21.6)	(20.0)

a. Represents the total money supply (M1) and quasi-money (time and savings deposits plus residents' foreign currency deposit).
b. Figures in parentheses denote the rate of change over the end of the previous year.
c. Represents the total currency in circulation and deposit money.
d. Represents money supply (M2) plus deposits of non-bank financial institutions.
Source: Bank of Korea.

Table 8. Interest rate on industrial paper (per cent per annum)

	End of						
	1978	1979	1980	1981	1982	1983	Apr. 1984
Industrial paper[a]							
With recourse	17.0	22.0	20.4	18.3	9.0	8.5	8.5
Without recourse	20.5	24.5	23.1	21.6	11.0	10.5	10.5
Commercial paper[b]	—	—	—	23.0	14.0	13.0	12.5
IFC's own paper[a]	17.0	19.5	16.4	15.7	8.0	8.0	8.0
Time deposits of DMBs[c]	15.0	15.0	14.8	14.4	7.6	7.6	6.0

a. Maturity: 90 days.
b. Maturity: 91 days.
c. Maturity: 3 months.
Source: Bank of Korea.

Therefore the most important thing, in terms of stabilizing prices, is to safeguard the independence of the Central Bank in the course of implementing monetary policy, because the Central Bank is relatively conservative in economic management and also regards the stability of the currency value as its primary objective. Traditionally, when the power of the Central Bank rather than that of the government has been strong, price stability has resulted. Thus the independence of the Central Bank from the government in the management of financial policy is considered one of the key factors for financial liberalization.

The Costs and Benefits of Financial Liberalization

When a government decides to undertake financial reforms, it should consider the possible consequences, both positive and negative, and implement the reform measures gradually.

In other words, it is necessary for the government to make a wide-ranging review of the complete liberalization process and draw up comprehensive long-term plans if the reform efforts are to have the intended effects.

In brief, we must review the potential demerits as well as the merits of financial liberalization, and prepare some contingent supplementary measures to maximize the positive aspects.

For example, we should keep in mind that the private ownership and operation of banks does not necessarily mean that banks will thenceforth be free from all controls and regulations. On the contrary, they might have to be subject to a new set of regulations and controls, because the government has to ensure that bank operations are compatible with the needs of the national economy. The areas which require a new system and regulations will include deposit insurance and the system of disclosure.

Table 9. Loans of deposit money banks by economic sector (percentages)

	End of			Apr. 1984
	1981	1982	1983	
Corporations	78.8	75.1	73.2	72.1
Large enterprises	(50.4)[a]	(48.6)	(44.7)	(44.4)
Small and medium enterprises	(49.6)	(51.4)	(55.3)	(55.6)
Individuals	19.0	21.8	23.7	25.0
Public entities and others	2.1	3.1	3.1	2.9
Total	100.0	100.0	100.0	100.0

a. Figures in parentheses denote the shares of the total loans to corporations.
Source: Bank of Korea.

The Relationship between the Real and the Financial Sector

It is conceivable that the development of the real economy cannot be achieved without a parallel and balanced development of the financial sector and vice versa, since both are closely interrelated.

Therefore, in principle, it is desirable to proceed with the internationalization of financial markets or the liberalization of capital markets only after we have achieved an equilibrium in the balance of payments; an industrial restructuring programme of the real sector should also proceed in line with the pace of financial liberalization.

Liberalization Initiated from inside the Financial Sector

Financial liberalization should naturally be initiated not from the outside, but from inside the financial community.

In the past, one of the characteristics of the Korean financial industry has been government-induced finance. Ironically, the financial liberalization currently under way is guided by the government, and thus it can be referred to as a government-led financial liberalization.

It is necessary that those engaged in the financial industry should have self-reliant and positive attitudes toward financial liberalization. As long as they lack these things, true liberalization will never be realized.

In any event, the point we would like to emphasize is that the government policy to readjust the scope of financial activities should focus on enhancing the efficiency of financial institutions, and minimizing the conflict of interests stemming from them, by encouraging bankers to take the lead in liberalization moves.

Table 10. Nominal and real interest rates

	GNP deflator (rate of change)	Curb market rate[a]		Yields on corporate bank		Yields on government bonds		Bank lending rate[b]	
		Nominal	Real[c]	Nominal	Real	Nominal	Real	Nominal	Real
1979									
1/4	20.1	44.0	23.9	26.1	5.9	23.4	3.3	19.1	−1.1
2/4	20.3	42.1	21.8	26.8	6.5	24.2	3.9	19.0	−1.3
3/4	21.4	40.7	10.3	26.9	5.5	25.5	4.1	19.0	−2.4
4/4	22.6	42.7	20.1	27.1	4.5	27.5	4.9	19.0	−3.6
Average	21.2	42.4	21.2	26.7	5.5	25.2	4.0	19.0	−2.2
1980									
1/4	25.5	50.8	25.3	30.5	5.0	30.3	4.8	24.3	−1.2
2/4	27.8	48.8	21.0	31.9	4.1	30.6	2.8	24.7	−3.1
3/4	23.7	42.5	18.8	29.7	6.0	27.9	4.2	23.7	—
4/4	25.9	37.7	11.8	28.1	2.2	26.2	0.3	20.8	−5.1
Average	25.6	45.0	19.4	30.1	4.5	28.8	3.2	23.4	−2.2
1981									
1/4	21.7	36.6	14.9	24.9	3.2	24.8	3.1	20.0	−1.7
2/4	17.0	35.2	18.2	22.8	5.8	22.2	5.2	20.0	3.0
3/4	17.1	33.8	16.7	22.9	5.8	21.8	5.7	20.0	2.9
4/4	10.5	35.4	24.9	27.0	16.5	25.5	15.0	19.0	8.5
Average	15.9	35.3	19.4	24.4	8.5	23.6	7.7	19.8	3.9

1982								
1/4	13.2	32.6	18.6	21.7	8.5	20.5	7.3	16.1	2.9
2/4	7.6	33.1	25.5	17.3	9.7	17.1	9.5	13.9	6.3
3/4	6.6	27.5	20.9	14.3	7.7	15.0	8.4	10.0	3.4
4/4	3.7	29.0	25.3	15.7	12.0	16.7	13.0	10.0	6.3
Average	7.1	30.6	23.5	17.3	10.2	17.3	10.2	12.5	5.4
1983								
1/4	4.8	24.1	19.3	14.9	10.1	14.4	9.6	10.0	5.2
2/4	2.2	27.5	25.3	14.0	11.8	13.5	11.3	10.0	7.8
3/4	2.2	26.8	24.6	14.0	11.8	13.4	11.2	10.0	7.8
4/4	2.9	24.7	21.8	14.2	11.2	13.8	10.9	10.0	7.1
Average	3.0	25.8	22.8	14.2	11.2	13.8	10.8	10.0	7.0
1984								
1/4	1.2	24.1	22.9	14.0	12.8	13.4	12.2	10.4	9.2
2/4	2.6	25.7	23.1	13.3	10.7	13.6	11.0	10.5	7.9
3/4	6.0	23.5	17.5	14.4	8.4	14.7	8.7	10.5	4.5
4/4	5.1	25.4	20.3	14.9	9.8	15.2	10.1	11.1	6.0
Average	3.9	24.7	20.7	14.2	10.2	14.2	10.2	10.6	6.6

a. Bank of Korea survey data.
b. Interest rate on bank loans up to one year.
c. Real interest rate = nominal interest rate – the rate of change of GNP deflator.

Source: Bank of Korea, *Economic Statistics Yearbook*.

Table 11. Return on total capital (percentages)

	Net profit (100 million won)		Return on total capital		
	Foreign bank branches	Domestic commercial banks	Foreign bank branches	Domestic commercial banks	World's 35 largest banks[a]
1978	130	480	1.4	0.7	0.62
1979	191 (67.3)[b]	551 (14.8)	1.6	0.6	0.64
1980	451 (111.7)	823 (106.8)	1.6	0.7	0.62
1981	583 (29.3)	850 (3.3)	1.7	0.5	0.6
1982	607 (4.1)	406 (−52.3)	1.4	0.2	0.57
1983	617 (1.6)	312 (−23.2)	1.3	0.2	0.67[c]
1984	775[d] (25.6)	791 (153.5)	1.59	0.3	—

a. Salomon Brothers, *Review of Bank Performance*, 1983.
b. Ratio of increase compared with the previous year.
c. World's 500 largest banks.
d. Includes profits from Japanese banks as of 1983 fiscal year.
Sources: Bank of Korea, Ministry of Finance.

Future Tasks

There are, however, a large number of difficult tasks to be completed before full financial liberalization is achieved. Some of those remaining include a liberalization of interest rates, an enhancement of the international banking capabilities of domestic financial institutions, and a restructuring of the institutional framework of the financial market for greater stability and efficiency.

Interest-rate Policy

In the organized financial market, virtually all interest rates are controlled by the monetary authorities. This rigid and arbitrary interest-rate determination tends to result in poor responses to changes in the macro-economic situation, an uneven growth pattern among different types of institutions, and an inefficient allocation of financial resources. Although the weak financial structure of Korean business firms imposes constraints, interest rates should gradually be liberalized and determined by market forces.

As inflationary expectations recede with continued price stability, the gap

Table 12. Trend of foreign bank branches' market share (percentages, as of the end of the individual year)

	1978	1979	1980	1981	1982	1983	1984
Loans	8.9	9.1	13.2	11.9	11.5	11.5	13.0
Korean currency	4.0	3.3	5.0	5.3	5.4	5.6	6.2
Foreign currency	38.4	39.2	51.5	50.1	50.3	56.9	74.3
Deposits received	1.4	1.0	1.5	1.7	2.4	2.1	2.2
Korean currency	1.3	1.0	1.3	1.3	1.1	1.3	1.4
Foreign currency	2.0	1.7	3.7	9.5	34.5	22.5	21.3

Source: Bank of Korea, *Monthly Statistics.*

between official and free market rates should continue to narrow. This will allow the government to liberalize interest rates on money market instruments and ease the regulation of bank interest rates. In addition, the government will broaden the range of bank lending rates, and set more flexible deposit rates.

Policy Loans

Commercial banks will gradually be freed from the obligation to make preferential policy loans, a burden which will be assumed by the Korea Development Bank and the government. The managerial autonomy of banking institutions will be further promoted, while management will become more responsible for operational performance.

An extensive review of the Korean financial system is currently under way to identify potential institutional innovations. The Korean government is committed to a more competitive and integrated financial market. There will be fewer limitations on the types of services offered so that specialization of financial services will emerge in response to the needs of the market rather than by legislation. At the same time, entry barriers will be lowered even further to promote competition.

Government ownership of some specialized banks will be handed over to the private sector; since activities of these banks are almost identical to those of commercial banks, there seems to be no reason for their existence. New financial markets for such instruments as BAs and RPs will be developed, and diverse financial services such as leasing, securities, consulting, and others utilizing advanced banking techniques will be promoted. As the money market deepens, short-term finance companies will be promoted as dealers or brokers.

Effects on Monetary Policy

Still in their early stages, Korean financial reforms have not posed any serious problems for monetary policy, even though there have been some signs of instability in money demand functions. Monetary authorities, therefore, do not feel

Table 13. Market share of foreign bank branches in each country (percentages)

	Republic of Korea (1984)	Japan (1983.3)	Taiwan (1982)	FRG (1982)	Singapore (1981)
Deposits	2.1	0.8	2.0	2.7	84.9
Loans	12.8	3.5	7.8	2.2	76.5
Total assets	8.5	4.3	6.7	2.1	83.4

Source: Bank of Korea.

any immediate need to change the current policy of announcing and managing the growth rate of total money stock as an intermediate target of monetary management.

The diversification of financial assets and the increased importance of non-bank financial institutions resulting from recent financial reforms have made it more difficult for monetary authorities to control monetary aggregates by increasing the degree of substitutability among financial assets. For instance, reduced entry barriers for non-bank financial institutions, which are highly competitive with banks, seem to have drained a considerable amount of deposits out of banking institutions; this situation has affected the stability of the relationship between broad money (M2) and economic activity. Moreover, the introduction of new deposits, such as household checking deposits, has blurred the traditional concept of narrowly defined money (M1) as cash plus demand deposits of banking institutions. In order to adjust to these drastic changes of financial environment, monetary authorities should create a new concept of money which reflects the proper relationship between money and real economic activities.

Degree of Internationalization

Another factor which might affect the prospects for financial reform is the importance of the external sector in the economy. As the economy grows in size and becomes more dependent on the external sector, there will be a growing demand from abroad to open the financial industry to the rest of the world. At the same time, there has been some domestic resistance to this opening.

Indeed, the Korean financial sector is far from being internationalized. Korea lacks the manpower and know-how necessary for international banking. We will therefore have to upgrade business capacity in this area by providing more training opportunities in the field of international finance and by utilizing experienced foreign advisers. In this regard, liberalization of the domestic financial market is an important step.

The entry of foreign banks and non-bank financial institutions, such as merchant banking corporations, will be allowed in order to help domestic financial institutions upgrade their international banking services. As more Korean busi-

Table 14. Financial market outlook (billion won, percentages)

	1984 end balance	Increasing rate	
		1984 (estimated)	1985 (forecast)
Money supply (M2)[a]	24,763	9.0	9.5
Money supply (M1)	6,886	6.0	8.5
Time and savings deposits	17,877	10.7	10.0
Transferable deposits and trust account	2,906	68.9	30.0
Non-bank deposits	22,737	28.1	21.5
Insurance	5,207	41.4	31.3
Short-term financing	7,021	27.7	15.7
Others	10,509	22.6	20.6
Securities	12,393	22.3	18.8
Bond	9,531	22.9	18.6
Stock	2,862	20.1	19.2
Total	56,820	18.2 (8,760)[b]	16.5 (9,370)

a. Includes residents' foreign currency deposits.
b. Figures in parentheses denote the amount of increase.
Source: Ministry of Finance.

ness firms develop into multinational corporations with major overseas operations and look to the international financial markets for their capital mobilization, Korean financial institutions, in competition with foreign institutions, will have to provide better services to meet the financing needs of these firms.

Korean financial institutions also need to be better represented in the international financial markets and to grow as an integrated part of those markets by offering diversified financing instruments to a wide range of customers.

To facilitate the internationalization of the financial sector, we must foster the foreign exchange market with a relaxation of exchange controls. As Korea's current account is expected to show structural equilibrium within three to four years, foreign exchange controls on services transactions are likely to be eased. The positive listing of designated currencies will be replaced by a negative list, and the limited futures market for foreign exchange will grow to be more broad-based.

Also to be relaxed are capital transactions in the financial sector. Transactions such as interest-rate swaps, long-term currency swaps, and foreign-currency-denominated factoring have already been introduced. Equity investment in financial institutions by other financial institutions and direct investment in equities of Korean business firms by foreign investors will be allowed in the coming years.

The Long-term Plan for Liberalization

Financial reform has been an integral part of structural reforms to improve allocative efficiency in the real sector. Since its scope and timing will have to be closely related to the real sector of the economy, the government should undertake a comprehensive review of the whole liberalization process to formulate a long-term plan. Pursuit of liberalization, of course, does not mean we can solve all of the problems in the financial sector. It is crucial, however, that the government carry out the liberalization policy in a rational and systematic manner.

Conclusion

Together with the modernization of the economy, the financial sector has continued to evolve through the diversification of financial services and functions of financial intermediaries. In comparison with the real sector of the Korean economy and the financial sector in foreign countries, they are both at a similar stage of development, yet Korea's financial sector still remains underdeveloped in many respects.

The financial reform of the early 1980s has contributed greatly to the efficiency of the financial industry. A series of policy measures, including the government decision to relinquish its equity holdings in commercial banks, the opening of newcomers' participation in the money market and mutual savings industry, and the lessening of market segmentation among financial institutions all demonstrate the government's will to carry out financial reform.

Although the domestic banking sector has maintained close links with international financial markets, Korea's financial institutions still seem to be inactive in international banking. Recently, the country's ever-increasing foreign debt has loomed as a serious problem facing the economy and has become a more important policy target.

In this context, the active participation of financial institutions in international financial markets is needed now more than ever. Advanced countries themselves are undergoing fundamental changes in the financial services industry. Such a trend has forced Korean policy-makers to review existing understandings and theories on the issues related to financial liberalization, including the role of money and finance and the effect of monetary policies.

There exists in Korea a dual financial structure: the organized and the unorganized market. The government has been trying to integrate these two markets by establishing short-term financial institutions and improving the performance of the long-term capital market. Banks are by far the most important financial institutions in Korea, and the government has tried partly to offset their inefficiency by developing non-banking financial institutions. These have made great strides in recent years, but the desired integration of the two markets has not fully materialized.

The future of Korea's financial services industry is bright, provided that the government continues to forge ahead with the current strategy, which is drastically different from that of the past. If the current thrust of economic policies is not maintained, the chances for a genuine financial liberalization will indeed be very slim, in spite of the privatization of banks. The liberalization of the financial services industry also requires the arrest of inflationary pressure, which is inherently inimical to its development.

After all, the ultimate goal of liberalization is to raise efficiency. This implies that great attention should be paid to giving banks a maximum degree of autonomy in their decision to lend or borrow, in contrast to deregulating the deposit interest rate. Finally, we would like to emphasize that the financial liberalization process should proceed gradually in parallel with deregulations on the real side.

Either the effectiveness of financial deregulation will be reduced, or the deregulation will eventually lead to regulation of another type, if non-market decisions continue to be a factor in investment, exports, and imports sectors.

Notes

1. H.T. Patrick, "Financial Development and Economic Growth in Underdeveloped Countries," *Economic Development in Cultural Change*, January 1966; E.S. Shaw, *Financial Deepening in Economic Development* (Oxford University Press, London, 1973).
2. R. McKinnon, *Money and Capital in Economic Development* (Brookings Institution, Washington, D.C., 1973).
3. The evidence often cited includes the low financial interrelations ratio, and the low ratio of financial assets held by business firms compared with the volume of sales of these firms. Furthermore, the scale of Korean commercial banks is very small; as a comparison, the scale of seven Korean commercial banks is said to be about the same as that of a small Japanese provincial bank.
4. The main purpose of the Presidential Emergency Decree in 1972 was to improve the financial structure of private loan-ridden corporations and to institutionalize these loans.
5. Financial regulation and change in the financial system are directly or indirectly related to the government objective of upgrading financial resources. See T.F. Cargill and G.G. Garcia, *Financial Deregulation and Monetary Control* (Hoover Institution Press, Stanford, Calif., 1982).
6. The maximum ownership of any single shareholder was limited to 8 per cent of the total shares (except for joint ventures or local banks) and guarantees and accep-

tances for a single beneficiary were restricted to a the maximum of 50 per cent of a bank's net worth.
7. Many monetarists believe that monetary control via a reserve requirement instrument is more efficient than control through the interest rate, yet there is relatively little historical experience on which to base a test of the proposition.
8. In November 1984, the range for bank lending rates was widened further, from 10 to 11.5 per cent.
9. When anonymous or fictitious names are used, income from financial assets was taxed 50 per cent higher during the 18-month period from July 1983 and 100 per cent higher from January 1985.
10. The government's long-term blueprint announced in 1981 envisaged four stages of development towards the complete opening of the market. They are: (1) indirect investment in Korean securities through special trusts for foreigners (up to 1984); (2) direct securities investment by foreigners on a limited basis (around 1985); (3) full-scale direct investment by foreigners and the issuance of local securities overseas (late 1980s); and (4) the full liberalization of capital movements (late 1980s).

Bibliography

Akhtar, M.A. *Financial Innovations and Their Implications for Monetary Policy: An International Perspective*. Bank for International Settlements, Basle, 1983.

Bank of England. "The Nature and Implications of Financial Innovation." Report of a conference of central bankers in London, 18–20 May 1983. *Quarterly Review* (Bank of England), September 1983, pp. 358–362.

Brown, Gilbert. *Korean Pricing Policies and Economic Development in the 1960s*. Johns Hopkins University Press, Baltimore, Md., 1973.

Cargill, T.F. *Money, the Financial System and Monetary Policy*. Prentice-Hall, Englewood Cliffs, N.J., forthcoming.

Cargill, T.F., and G.G. Garcia *Financial Deregulation and Monetary Control*. Hoover Institution Press, Stanford, Calif., 1982.

———. *Financial Reform in the 1980s*. Hoover Institution Press, Stanford, Calif., forthcoming.

Cole, David, and Yung Chul Park. *Financial Development in Korea, 1945–1978*. Council on East-Asian Studies, Harvard University, Cambridge, Mass., 1983.

Federal Reserve Bank of St. Louis. *Financial Innovations: Their Impact on Monetary Policy and Financial Markets*. Proceedings of a conference held at the Federal Reserve Bank of St. Louis, 1–2 October 1982. Kluwer Nijhoff Publishing, Boston, 1984.

Fraser, D.R., and P.S. Rose, eds. *Financial Institutions and Markets in a Changing World*. Business Publications, Plano, Tex., 1984.

Greenbaum, S.I., and Bryon Higgins. "Financial Innovation." In: G.J. Benston, ed., *Financial Services: The Changing Institutions and Government Policy*. The American Assembly, Columbia University, N.Y., 1983.

Hester, Donald D. "Innovations and Monetary Control." *Brookings Papers on Economic Activity*, 1 (1981).

Howard, D.H., and Karen H. Johnson. "Financial Innovation, Deregulation and

Monetary Policy: The Foreign Experience." *Interest Rate Deregulation and Monetary Policy*. Federal Reserve Bank of San Francisco, San Francisco, 1982.

Kim Joong-Woong. "Financial Policy in Transition." *The Korea Newsletter*, 23 June 1984, pp.5–6.

McCarthy, F. Ward, Jr. "The Evolution of Bank Regulatory Structure: A Reappraisal." *Economic Review*, March–April 1984, pp.3–21.

McKinnon, Ronald I. *Money and Capital in Economic Development*, Brookings Institution, Washington, D.C., 1973.

Morris, Frank E. "Do the Monetary Aggregates Have a Future as Targets of Federal Reserve Policy?" *New England Economic Review* (Federal Reserve Bank of Boston), March–April 1982, pp.5–14.

Niehans, Jurg. "Innovation in Monetary Policy: Challenge and Response." *Journal of Bank Research*, March 1982, pp.9–28.

Shaw, Edward S. *Financial Deepening in Economic Development*. Oxford University Press, London, 1973.

Porter, R.D., and E.K. Offenbacher. "Financial Innovations and Measurement of Monetary Aggregates." *Financial Innovations: Their Impact on Monetary Policy and Financial Markets*. Proceedings of a conference held at the Federal Reserve Bank of St. Louis, 1–2 October 1982. Kluwer Nijhoff Publishing, Boston, 1984.

Patrick, Hugh T. "Financial Development and Economic Growth in Underdeveloped Countries." *Economic Development in Cultural Change*, vol. 14, no. 2, January 1966.

Simpson, T.D., and P.M. Parkinson. "Some Implications of Financial Innovations in the United States." Staff Studies Series, no. 139. Board of Governors of the Federal Reserve System, Washington, D.C., 1984.

FINANCIAL OPENING IN MALAYSIA

Andrew Sheng

Introduction

In recent years, the role of financial markets in economic development has received increasing interest from development economists. The early studies viewed financial markets as having a more or less passive role in development, with some studies concentrating on the exploitative effects of high interest rates on rural credits in unorganized money markets, and the dual development between these unorganized markets and efficient, modern, urban-based financial systems. The pioneering work of Goldsmith called attention to the importance of the development of financial markets in developing countries, and was followed by that of Gurley, Shaw, McKinnon, and Patrick. These studies suggested that the process of financial liberalization and encouragement of efficient markets through "deepening" and elimination of "fragmentation" of markets could improve the process of mobilization of savings, as well as the efficiency of investments. This would also improve resource allocation and raise the yield on capital and level of output. Patrick's view that "demand-following" financial development – through the creation of modern financial institutions, instruments, and financial services as a consequence of developments in the real economy – may be too passive has led to the suggestion that the lack of financial

---

The views expressed in this paper are solely those of the author and do not necessarily reflect those of the Bank Negara Malaysia.

services inhibits the process of growth. Alternatively, planners should move towards a "supply-leading" phenomenon, to the deliberate creation of financial institutions, instruments, and services in advance of demand, to achieve the twin objectives of transferring resources from traditional low-growth sectors to modern and high-yielding areas and to stimulate and promote entrepreneurial development in such sectors.

A synthesis of the key roles of the financial sector seems to suggest that finance plays four major functions in development. First, the establishment of a stable medium of exchange improves the transactions mechanism in the economy. Secondly, an efficient financial system changes the maturity profile of savings in an economy and maturity profile of resource applications, such as investments or lending to enterprises, thus distributing and reducing risks for both small savers and entrepreneurs. Thirdly, an efficient financial system creates the resource requirements in the form of financial instruments in quantities and qualities tailored to the needs of savers and investors, to ensure that savings go to areas where investments yield the highest return for output growth. Fourthly, the financial system is the mechanism through which planners can stabilize the economy by appropriate monetary and financial policies, in particular, via the money and foreign exchange markets. In other words, the financial markets are the mechanisms through which governments can influence the level of interest rates and the exchange rate in an economy. Developing countries should, therefore, try to improve the efficiency of their financial systems in order to speed up the process of development.

However, the activist role of governments in financial development in recent years, particularly through the creation of development banks, credit controls, and ceilings on interest rates in developing countries, has led to what McKinnon calls the "repression" of financial development and the fragmentation of markets, causing considerable resource distortion. In particular, the record of the LDCs' development banks in promoting sectoral growth has been mixed, and the administrative inefficiencies and misallocation of resources that have led to high external debt financing requirements suggest that a liberalization process could remove the bottlenecks of development and open the way to true supply-leading financial reform. In particular, with the onset of the twin oil shocks, the growing internationalization of banks, financial innovation, and improvements in banking technology in the late 1970s and early 1980s, financial policy planners in LDCs found thay they could no longer insulate their domestic economies and financial markets from the effects of external inflation, high interest rates, and volatile capital flows that distorted the relative calm of economic and financial management in the 1960s. The experience of Malaysia during this period is one that could give some useful insights into the behaviour of the financial markets in a small open economy in this interdependent world.

The Malaysian Financial System

The Malaysian financial system is relatively sophisticated for a country at its present stage of growth. The ratio of broad money (M2) to GNP increased from 36 per cent in 1974 to 59 per cent in 1983, the second highest in ASEAN after Singapore, higher than the Republic of Korea (36 per cent) and comparable to Japan in the 1950s. At the end of 1984, the total gross assets of the financial system amounted to M$147.5 billion or slightly more than twice the size of GNP. The assets of the banking system, comprising the Central Bank, commercial banks, and other monetary institutions, accounted for three-quarters of the total assets of the financial system; the commercial banks alone accounted for 45.5 per cent of those assets. Today, the whole system comprises a Central Bank at the apex, which supervises and regulates 39 commercial banks, 41 finance companies, 12 merchant banks, 5 discount houses, and 7 money and foreign exchange brokers. The other five major groups of financial institutions comprise the development banks, the savings institutions (including the national post office savings bank), the provident and pension funds, the insurance companies and other specialist financial institutions.

Concerning the distribution of the financial institutional network throughout the country, the commercial banks have a network of 608 offices, while the financial companies have 303 branches. With 23,000 population per branch office, Malaysia is relatively well served by banking offices. The sophistication of the financial system can also be seen from the lengthening in maturity of financial instruments in recent years. By the end of 1984, deposits of more than one year in maturity with the financial system accounted for 21.5 per cent of total deposits, compared with less than 1 per cent at the end of 1970 and 11.5 per cent in 1980. At the same time, loans and advances of more than four years maturity more than doubled from 16.6 per cent of total loans in 1970 to 38.4 per cent at the end of 1982. It is not uncommon for home purchasers to obtain housing mortgages of 20 to 25 years, while industrial and agricultural loans of 7 to 10 years are relatively easy to obtain.

The degree of financial deepening is also illustrated by the fact that domestic savings in financial assets accounted for 78 per cent of private saving in 1979–1983, compared with 49 per cent in 1975–1978.

Malaysia has also a sizeable capital market, with an active government securities market and corporate stock exchange. Comprising nealy 280 quoted companies with a par value of M$20 billion and a market capitalization of M$70.3 billion, the Kuala Lumpur Stock Exchange (KLSE) is one of the most active capital markets amongst developing countries. The market is closely linked to the neighbouring Singapore Stock Exchange (SES). Malaysian companies account for 53 per cent of all shares traded on the SES and average annual turnover in the early 1980s amounted to M$8 billion, compared with M$14 billion in the SES. In 1984, the Kuala Lumpur capital market raised

M$5.6 billion in gross securities issues, comprising M$3.2 billion in government securities and M$2.4 billion (42.7 per cent) in private corporate securities.

Phases of Financial Development

The development of the Malaysian financial system may be divided into four phases: the pre-Independence period to 1957; financial development, 1957–1967; adaptation and change, 1968–1978; and the post-interest rate liberalization period since 1978.

In the pre-Independence period, the financial system was characterized by its colonial heritage, with a Currency Board monetary structure, a banking system dominated by the colonial trading banks, and a post office savings bank. A landmark institutional development was the establishment of the Employees' Provident Fund (EPF) in 1951, to provide provident fund benefits for low-income groups. This institution today alone accounts for half of the federal government's domestic debt and has been a major source of forced savings which has successfully funded Malaysia's infrastructure development.

After independence in 1957, the government rapidly proceeded to establish the Central Bank (1959), the first development bank (1960), the local stock market (1960), and the incorporation of domestic banks. At the end of 1959, only 8 out of 26 banks in Malaysia were locally incorporated. From the outset, the Central Bank took an active supply-leading role in financial innovation and development by creating financial instruments, institutions, and markets where possible in advance of demand. These innovations comprised not only the design of the necessary legislation and institutional structure of new markets, but also the staffing and training of personnel for such institutions or markets, often with the assistance or participation of foreign banks and institutions. In 1963, the first discount house was established to create the inter-bank market and in 1964 a treasury bills and government securities market was created with the Central Bank as market-maker. Following the separation of Singapore from Malaysia in 1965, the joint Currency Board system was dismantled and the Central Bank took over the currency issue function in 1967.

The second decade of independence saw rapid changes in the Malaysian economy, with the onset of the first oil shock and rapid growth arising from the commodity boom in 1972–1973. As branch banking grew rapidly with monetization during this period, more major financial institutions came into play. The deposit-taking finance companies were brought under the supervision of the Central Bank in 1969, followed by the deliberate creation of the first merchant bank in 1970. In 1973, in the wake of the separation of common financial facilities and institutions from Singapore and the floating of the Malaysian ringgit in June, a series of major financial reforms took place, including the following:

1. Establishment of an independent foreign-exchange market through licensing of six foreign-exchange and money brokers.
2. Establishment of a separate Kuala Lumpur Stock Exchange.
3. Introduction of a non-discriminatory and liberal exchange-control system.

Since 1973, developments in the Malaysian financial scene have accelerated, growing in breadth as well as in depth. With growing competition amongst the financial institutions, savings mobilization grew rapidly with the rapid rise in real income as a result of rising commodity prices and exports. By 1978, the Central Bank had prepared the ground for major changes in financial reforms in order to enhance the effectiveness of its monetary control tools and to improve competitiveness in the banking system.

Firstly, the system of administered interest rate for commercial banks was abolished in October 1978, and the banks were allowed to determine freely their deposits and lending rates. However, lending guidelines through special groups or priority sectors and ceiling rates remained basically unchanged. Secondly, the liquidity requirements of the commercial banks and finance companies were restructured. Thirdly, two new monetary instruments, namely bankers' acceptances and negotiable certificates of deposits, were introduced.

The period since 1978, following the liberalization of the interest rate regime, has seen rapid advances in financial deepening. Growth in NCDs amounted to M$3.2 billion at the end of 1984 while bankers' acceptances (BAs) amounted to M$2.2 billion. In 1980, Central Bank Certificates were introduced to augment monetary instruments for liquidity control. In terms of institutional development, a Kuala Lumpur Commodity Exchange was established in 1980 and two new development banks were created to provide long-term loans in specialized and sectoral fields. To meet the needs of the Muslim population, an Islamic Bank was created in 1983.

Financial Development and Real Growth

The rapid financial deepening in Malaysia has coincided with significant achievements in real growth in the economy since Independence. In the 1960s, Malaysia achieved an average annual real growth of over 5 per cent with inflation of less than 1 per cent. By the 1970s, real growth had accelerated with improvements in the terms of trade to average 8 per cent annually, while inflation deteriorated to average about 5 per cent annually, due partly to supply bottlenecks and imported international inflation. The growth in the real economy was particularly strong in the second half of the 1970s, when the sharp improvement in the terms of trade for non-oil commodities, and the advent of oil and gas production, brought about large surpluses in the balance of payments, substantial capital inflows, and the commencement of large public sector investments in infrastructure and heavy industries.

The success of the 1960s and 1970s in growth with low inflation could be attributable to the following key factors:

1. The high domestic savings rate of about 30 per cent of GNP, a significant proportion of which was captive in the form of provident fund contributions. These funded the infrastructural development programme of the government, which had a fiscal deficit averaging 5 per cent of GNP in the 1960s and 1970s, funded mainly from the non-inflationary sources.
2. A strong commitment to financial discipline in the fiscal budget, with high levels of public sector savings – the federal government runs a current account surplus for every year since Independence and funds about one-quarter of development spending from internal savings – and non-resort to Central Bank financing. Borrowing by state governments and public enterprises was strictly controlled, and external borrowing was limited to residual financing of development expenditure.
3. The open nature of the economy and its outward-looking development approach. Exports account for over 55 per cent of GNP, one of the highest for developing countries, while the degree of protection has remained relatively low. For most of the first two decades of independence, the government undertook a cautious import-substituting strategy, switching to more export-oriented industries through free-trade zone development in the 1970s. At the same time, considerable resources were devoted to agricultural research and diversification, resulting in world-leading production of rubber, palm oil, pepper, timber and, by the end of the 1980s, cocoa. The country also maintained a liberal foreign exchange control regime, which allowed almost complete freedom of movement in funds for both the current and capital account.
4. Appropriate interest-rate and exchange-rate policies. In line with the openness of the economy, the Central Bank maintained interest rates at positive real levels throughout the period, except selected years in early 1980–1981. At the same time, with the exchange rate initially pegged to the pound sterling and eventually against a composite basket of currencies of the country's major trading partners in 1975, the exchange rate was kept relatively stable and in line with the competitiveness of the economy.

Recent Developments and Resource Constraints

In the early 1980s, as in all developing countries, the international recession brought about by the last oil crisis and ensuing tight monetary policies of the industrial countries resulted in high international interest rates and a severe deterioration in the terms of trade. To maintain the growth momentum, the government initially adopted counter-cyclical policies and embarked on heavy investments, particularly in infrastructure and some basic industries, in order to take advantage of oil resources wealth availability and the need to maintain growth. Funded by the onset of oil production to over 450,000 barrels per day, the national gross investment rate increased to as high as 36 per cent of GNP in 1982, while the savings rate fell to 25 per cent of GNP, resulting in 1982, for the first time, in large current account deficits of 11 per cent of GNP, which were funded mainly by resorting to external borrowing. Since the country maintained high foreign exchange reserves and had low external debt, it obtained very favourable terms in the international markets. Nevertheless, the external debt increased to US$15 billion by the end of 1984, although the debt service ratio remained modest at 11 per cent of GNP, while the debt/export ratio was modest at about 90 per cent. Since public sector expenditure was running at a high level of about 40 per cent of GNP, and the fiscal deficit worsened to 19 per cent of GNP in 1982, the government took significant austerity measures by cutting expenditure, subsidies, and low priority investments sharply. With the recovery in world trade in 1983–1984, and with exports rising by an average of 16 per cent annually, the twin deficits were cut significantly, with the fiscal deficit reduced to 10 per cent of GNP and the current account deficit to 4.1 per cent of GNP by 1984.

Given the rising debt level and its servicing implications, as well as the expected difficult trading conditions in the rest of the 1980s, the government reviewed its own development and financing strategies. In the development field, in addition to increasing efforts to cut the fiscal deficits and improve public sector efficiency, an attempt would be made to increase agricultural output and private sector investment in resource-based industries. In addition, a complete review of the financial system was undertaken to ensure that resources would be available to fund the continued high level of investments, without resort to substantial external financing. This strategy called for further liberalization of the financial markets and deregulation.

Financial Reform and Liberalization

As mentioned earlier, Malaysia has already one of the more sophisticated financial systems amongst developing countries at its stage of development. Nevertheless, several institutional rigidities remain to inhibit completely free competition. In the late 1970s and early 1980s, the fierce competition for deposits arising from the lowering of the savings rate and rising investments caused severe competition and changes amongst the various financial institutions, with significant implications for monetary control and institutional regulation and supervision. With greater interest rate flexibility, the finance companies and other non-bank financial institutions grew in breadth and depth, resulting in the blurring of institutional differences. This phenomenon, also caused partly by changes in banking technology as well as greater monetization and interest-rate sensitivity, had far-reaching implications for the economy.

A review of these implications, which is at present still incomplete and under study, showed the following:
1. Given the high gross level of domestic savings in the economy, the possibility of major institutional changes to increase gross savings significantly remain limited in scope. Econometric studies showed that interest-rate changes had limited effects on the level of national savings, these being influenced more significantly by the growth in national income and the level of income.
2. Changes in the interest-rate structure would have a major impact on the efficiency of investments and on the allocation of resources. With an open economy and liberal exchange control, the level of interest rates would have to be sufficiently high to prevent capital outflows due to interest differentials.
3. Financial liberalization would have to be concentrated on removing the existing constraints towards full and free competition in the financial system. This involves the gradual move towards market-oriented interest rates on government securities, as well as the restructuring of credit controls, without distorting the efficient flow of resources to areas for socio-economic policy reasons.
4. The ultimate goal still remains to improve the efficiency and productivity of investments, through the removal of constraints in private sector investments in both manufacturing and agriculture, while streamlining the pace of public sector infrastructural development.

Conclusion

The experience of Malaysia suggests that financial development depends very much on the imposition of appropriate financial discipline, which remains the cornerstone of key monetary and fiscal policies. With broadly correct interest-rate and exchange-rate levels, and a fundamentally open economy, the process of private investments and efficiency in production can proceed to provide the necessary growth for the broader goals of income distribution and raising the standard of living of the populace, without high inflation. A supply-leading financial policy helps lay the groundwork for good financial deepening, but is not sufficient to secure good growth. What is vital in the long run is the need to ensure correct interest rates and exchange rate for the economy, free from policy-induced or structural distortions, which would allow the efficient allocation of resources towards investments. The correct and appropriate mix of investments, however, is beyond the scope of this paper. What is clear is that financial reform can help in leading development, but may open the economy more to international shocks. Policy-makers would therefore have to remove not only obstacles to financial "repression," but also those to investment "repression," in order to achieve the broader aims of growth with stability and equity.

ECONOMIC GROWTH AND REGULATION OF FINANCIAL MARKETS: THE JAPANESE EXPERIENCE DURING THE POST-WAR HIGH-GROWTH PERIOD

Juro Teranishi

Introduction

During the period 1953–1972, Japan experienced an exceedingly high rate of economic growth. The annual growth rate of real GNP during this twenty years was 9.2 per cent, exceptionally high when compared to the growth rate of 3.7 per cent for 1910–1938 and 3.9 per cent for 1973–1981. The change in industrial structure was also drastic. The share of the primary sector in GDP was 36.4 per cent in 1910 and 18.3 per cent in 1938, but fell to 6.1 per cent in 1970. The annual decrease in the labour force in the primary sector from 1950 to 1965 was 36,500 persons, six times as high as during the 1910–1940 period, when it numbered 6,200 persons. It was really a turbulent period.

What was particularly interesting for researchers in money and banking was that this rapid economic growth had taken place within a highly regulated financial framework. The time deposit rate was regulated, entry into the bond issue market was regulated, international asset transactions were virtually forbidden, and so on. In the pre-war period, the financial system was essentially free, in terms of both interest-rate movements and entry into financial markets. After about 1975, the financial markets were rapidly deregulated and

---

This is the revised version of the paper presented at the United Nations University Conference on Financial Liberalization and the Internal Structure of Capital Markets, held on 22–24 April 1985 in Tokyo. The author is very appreciative of the comments made by the participants in the conference, and is also greatly indebted to Kermit Schoenholtz of Yale University for his comments and suggestions.

liberalized, domestically as well as internationally, and therefore the system that prevailed during the period of high economic growth was rather an exceptional one.[1] Moreover, somewhat surprisingly for those who are familiar with McKinnon–Shaw hypotheses, the level of financial intermediation was quite high in spite of the deposit rate regulation. According to McKinnon's recent paper, the average M_2/GNP ratio for Japan from 1960 to 1975 was 0.832, which was very high when compared with 0.184 for four Latin American countries, 0.247 for four Asian countries, 0.585 for five industrialized countries, and 0.447 for four rapidly growing countries.[2] How was the high level of intermediation reconciled with strict regulations? This question will be dealt with below, under "Four Characteristics of the Financial System," as one of the four basic characteristics of the financial system during the period: (a) the high degree of intermediation and the underdevelopment of the securities market; (b) the coexistence and the division of labour between government finance and private financial institutions; (c) the accommodating stance of monetary policy; and (d) the insulation from the international financial market. These four characteristics are interrelated, since they have the following three conditions as a common background: a low level and equitable distribution of financial wealth; a high level of profits and informational needs within a rapid growth process; and favourable world trade conditions. It will be shown that these three conditions, together with compatible governmental regulations, were mainly responsible for the development of the four characteristics noted above.

The financial system during the high-growth period has often been called the "artificial low interest-rate policy." This suggests that there was a widespread belief that the policy of regulating financial markets has growth-promoting effects. This view has been criticized by Horiuchi in recent articles,[3] but the evidence remains inconclusive. "The Financial System and Economic Growth," below, gives an analysis of the possible effects of regulation on economic growth with regard to its macro-, micro-, and two-sector aspects: the effect on the overall cost of capital (macro-aspect) and the question of allocational efficiency under conditions of uncertainty (micro-aspect) are both considered, while from the viewpoint of two-sector development theory the role of adjustment funds in the change of industrial structure is discussed. Although the result obtained is quite inconclusive, it is hoped that our analysis will cast some light on the further study of this important question.

The following section classifies the financial markets and gives a brief account of the nature of regulations.

Regulation of the Financial System

Let us, for convenience, classify the various financial markets in the following manner.

1. Indirect finance
    - private financial system: deposit market* and loan market;
    - government financial system: deposit (postal savings) market* and loan market.*
2. Direct finance
    - direct lendings (including curb market);
    - securities market: bond (issue,* resale*) market, equity (issue, resale*) market and short-term money market.*

This classification is schematic and not intended to be precise. For example, the private financial system is mainly composed of commercial banking, but it also includes other financial intermediaries such as trust banks, insurance companies, and various institutions for small business or agriculture. Although the main source of funds for government finance is postal savings, funds are also obtained from the government annuity system and through flotation of government guaranteed bonds.

The markets with an asterisk above indicate the existence of some sort of (effective) official regulation. Within the intricate structure of implicit and explicit regulations which prevailed during the high-growth period, three kinds were important.

1. Regulation of interest rates
    - deposit rate (including postal savings rate);
    - issue rate of bonds;
    - official discount rate;
    - loan rate of government financial system.
2. Regulation of entry into the markets
    - bond issuance market;
    - short-term money market;
    - listing on stock-exchange market.
3. Regulation of international capital movements
    - inflow of funds;
    - outflow of funds.

The state of interest-rate regulation can be seen from table 1. Both the time deposit rate and the official discount rate were consistently lower than the call rates, a typical free-market rate in Japan.[4] (Note that there are no such clear-cut differences for the US.) Since the official discount rate was low compared to the market rate, Bank of Japan loans were rationed to banks, city banks being exclusive recipients, and have been a major tool of base money control. The continual dependence of the banking sector on Bank of Japan loans implies a loan supply in excess of available deposits. This feature of bank behaviour was called *overloan*. Since deposits were a cheaper source of funds for banks than money market resources, the supply function of deposits was horizontal at the regulated rate, and banks accepted whatever amount of deposits asset holders wanted to hold. The licensing policy of branch offices by the Ministry of Finance

Table 1. Interest rates in Japan and US

	1953–57	1958–62	1963–67	1968–72	1973–77
Japan					
Official discount rate	6.9	7.1	5.9	5.4	7.1
Call money rate	8.7	9.6	7.4	7.0	8.6
Time deposit rate	5.1	5.3	5.0	5.0	6.0
Rate of return on newly issued corporate bonds	n.a.	7.7	7.4	7.7	8.6
Market rate of return on corporate bonds	n.a.	8.4	8.1	8.0	8.8
United States					
Official discount rate	2.4	3.1	4.2	5.2	6.5
TB rate	2.1	2.7	4.0	5.4	6.2
Time deposit rate	2.6	3.2	5.0	5.0	5.5

Sources: A. Horiuchi (note 3); J. Teranishi (note 1), fig. 8-2.

was quite effective in controlling the market (deposit) shares among banking institutions. Throughout the period the licensing was less generous to city banks than to other banks, such as local banks, Sogo banks, Sinkin banks, and financial intermediaries for agriculture. On the other hand, these other banks were subject to various kinds of specialization requirements, either legal or *de facto*, with respect to their loan supply, whereas there was no such regulation for city banks, which were faced with a vast demand for loans, mainly from large business firms. As a result of this, the loan/deposits ratio of city banks was consistently higher than that of other banks, and, on the inter-bank money market (call market), city banks were borrowers and the other banks lenders. This phenomenon was called the imbalance of bank liquidity.[5]

The price of newly issued bonds was kept artificially higher than the prevailing price on the market, providing implicit subsidies to the issuer. This had two consequences. First, there arose excess supply in the market for bond issue and inevitable rationing by the committee, under governmental guidance. Corporate bond issuance was allowed only to firms in basic industries, such as electricity, with a considerable net worth. Second, bonds purchased by syndicated banks through assignment were rarely resold in the market since doing so would entail losses to the banks. Consequently, the resale market for bonds remained extremely underdeveloped throughout the high-growth period. The supply of funds by the government finance system is called the Fiscal Investments and Loans Programme (*zaisei-toyushi*), and rates of interest are kept at preferential low levels.

On the other hand, listing on the major stock exchanges was permitted only to relatively stable and large firms. The short-term money market was closed to non-bank economic units until the early 1970s, when the *gensaki* market (bond transactions with repurchase agreements) developed through spontaneous transactions between security companies and business firms. International capital transactions by both banks and non-financial institutions were virtually forbidden during this period. Mainly owing to the strict control of foreign exchange dealings, residents were unable to transact in foreign-currency-denominated assets or debts, and non-residents were unable to hold yen-denominated assets or debts.

Four Characteristics of the Financial System

The main characteristics of the financial system during the high-growth period can be summarized by the following four points: (a) a high level of intermediation and underdevelopment of the securities market; (b) the coexistence of and division of labour between government finance and private financial institutions; (c) an accommodating stance as regards monetary policy; and (d) insulation from the international capital markets.

The High Level of Financial Intermediation

The high level of financial intermediation in post-war Japan can be traced with the help of table 2, which shows the composition of sources of private-sector finances. It can be seen that a major proportion of outside funds was raised via the indirect financial system, and only a small proportion through issuance of bonds and equities. Going back to our schematic representation of the financial

Table 2. Source of finance in private industries

			Outside funds				
Year	Inside funds	Outside funds (total)	Borrowings from private finance institutions	Borrowings from government finance	Bond	Equity	Total
1956–60	42.7	57.3	41.8	4.7	2.7	8.1	100.0
1961–65	41.0	59.0	44.1	4.1	2.6	8.2	100.0
1966–70	49.2	50.8	41.3	4.6	1.6	3.4	100.0
1971–75	41.4	58.6	48.0	5.1	2.3	3.3	100.0

Source: *Showa-zaiseishi (Shusen kara Kowa made)*, vol. 19 (Toyokeizai-Shinposha, Tokyo, 1968), pp. 462–463.

Table 3. Composition of sources of borrowings by types of lenders for manufacturing by size of firms (percentages)

Size of firms by total assets (hundred yen)	1932[a]		Size of firms by number of employees	1957			
	Modern financing	Traditional financing		Modern financing (total)	Modern financing — Private	Modern financing — Government	Traditional financing
−1	11.6	88.4	1–3	65.7	56.0	9.7	34.3
1–5	12.5	87.5	4–9	74.8	65.0	9.8	25.2
5–10	15.1	84.9	10–19	83.2	73.7	9.5	16.8
10–20	19.1	80.9	20–29	85.5	76.7	8.8	14.5
20–50	28.9	71.1	30–49	86.8	78.7	8.1	13.2
50–100	36.3	62.7	50–99	86.7	79.4	7.3	13.3
100–500	42.5	57.5	100–199	88.8	82.9	5.9	11.2
500–1,000	39.3	60.7	200–299	86.4	82.6	3.8	13.6
1,000–5,000	52.7	47.3	300–499	91.1	89.5	1.6	8.9
5,000–	63.3	36.7	500–999	87.5	85.8	1.7	12.5
			1,000–	92.4	89.9	2.5	7.6
Total	60.8	39.2	Total	89.7	85.9	3.8	10.3

a. 1932 data for manufacturing firms in Tokyo and Kobe city.
Source: J. Teranishi (note 1), tables 6–21 and 6–24.

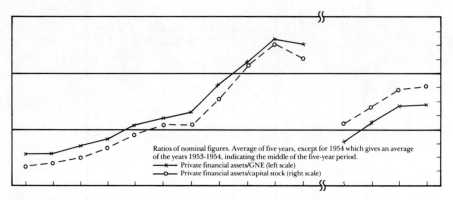

Fig. 1. Ratios of private financial assets to GNE and capital stock (after J. Teranishi [see note 1], p. 3).

system, this high intermediation ratio implies two things: a very limited role for direct lendings and an underdeveloped securities market.

The role of direct lendings, a typical method of traditional financing, decreased considerably after the Second World War. As table 3 shows, the share of traditional financing decreased from 39.2 per cent in 1932 to 10.3 per cent in 1957. After the war, even the smallest firm obtained 65.7 per cent of funds through the modern financial system. One of the reasons for this can be found in drastic changes in the level and distribution of asset holdings due to hyperinflation during and immediately after the war. From 1940 to 1950, every index of prices rose 100 to 300 times. As a consequence, private financial assets accumulated through the pre-war period lost most of their value. The ratio of private financial assets to GNE fell from 2.49 in 1936–1940 to 0.79 in 1953–1955 (fig. 1). Moreover, asset distribution became more equitable during the process. Small farmers were emancipated from the heavy burden of pre-war agricultural debt, while landlords and other wealthy classes virtually disappeared partly owing to inflation and partly to the land reform and the *zaibatsu* dissolution. Since traditional financial methods were essentially means of financing the poor by the rich, it was inevitable that the above-mentioned changes in the immediate post-war period reduced the role of such direct financing considerably. Another reason for the decreased share of direct lendings can be found in the informational needs that arose along with rapid growth. As we noted at the outset, this was a period of drastic changes in the industrial structure, and with a rapid emergence of new firms and products. As a consequence, the information available to individual asset holders was quite limited and inadequate for the efficient allocation of assets. In other words, direct lending could not keep pace with the massive informational needs, owing to the rapidly changing environment.

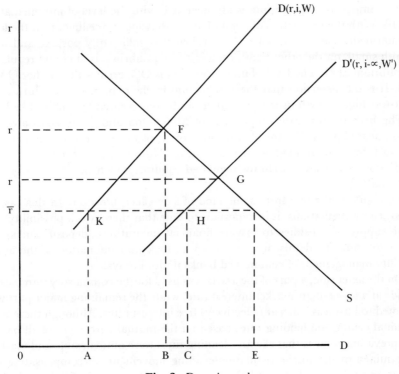

Fig. 2. Deposit market

It is easy to understand why the securities market remained underdeveloped. One reason was, obviously, the governmental regulations outlined above. Another reason lies in the reduced level of asset accumulation. At low levels of total wealth, asset holders (with utility function characterized by decreasing absolute risk aversion) become less inclined to take risks and hence reduce their investments in risky assets.

The reduced role of traditional financing, the underdevelopment of the security market, and the limited access to foreign asset holdings were the three major reasons for the high level of financial intermediation. In other words, since the availability of alternative assets was limited, people had no choice but to hold deposits in spite of the interest rate regulations on them. This situation is shown in figure 2. S and D are supply and demand functions of deposits, r and i represent rates of interest on deposits and on other assets (securities, direct lendings, and foreign assets) respectively, W and W' (<W) represent total wealth (deposits, other assets plus base money), and α shows the transaction costs of holding alternative assets (information cost plus imputed cost of decreased availability due to regulations). When α is zero and total wealth is

large, competitive equilibrium is attained at F with the level of intermediation at OB. With the imposition of regulations ($\bar{r}$ showing the ceiling rate), the level of intermediation is reduced to OA. When α is sufficiently positive and total wealth small, on the other hand, competitive equilibrium is at G and regulated equilibrium at H. The level of intermediation is OC, greater than either OA or OB. It is our assertion that the equilibrium in the deposit market during the post-war high-growth process in Japan can be represented by a point like H.

The high deposit rate proposition of McKinnon and Shaw advocates the attainment of point F by the liberalization of deposit rates (or of a point between K and F by maintaining high deposit rates).[6] However, at F, the level of intermediation can be lower than the regulated equilibrium whenever α is large and W' small.[7]

No regulation is free from some kind of side-effect, however. In this case of deposit-rate regulations, it is important to note that there was a possibility of a huge supply of subsidies to private financial institutions (deposit banks). In order to understand this, it is necessary to look into the nature of the asset-liability management of commercial banks under the system.

On the asset side, a part of the assets was used for the compulsory purchase of bonds at a lower-than-market interest rate, while the remaining major part was devoted to business loans at (effectively) free interest rates. Although there were nominal ceilings on lending rates based on the mutual agreement of all banks, the prevailing view is that effective lending rates were quite flexible and market-determined owing to the regulation-evading movements of compensating deposits and collateral.[8] On the liability side, a small part of the funds was raised in the form of Bank of Japan credits at a preferential interest rate, while a major proportion of liabilities was accounted for by private deposits at regulated rates. Bank of Japan credits took the form of either rediscounting commercial bills or the repurchase of bonds from banks at a preferential price (*riron kakaku*). In the latter case, banks were able to liquidate one-year-old bonds without loss.

With the framework of this system there are three kinds of implicit subsidies and taxation falling on banks: (1) implicit subsidies due to Bank of Japan borrowings = (call rate − official discount rate) × Bank of Japan borrowings; (2) implicit taxation due to bond holdings = (market rate − issue rate) × bond holdings; and (3) implicit subsidies due to deposit rate regulation = (call rate (1 − reserve ratio) − deposit rate) × deposits. Table 4 reports estimates of 1–3 for city banks (a major category of commercial bank). Estimates of 1 and 2 are rather small in magnitude, and both seem to offset each other. Estimates for 3, on the other hand, are quite large,[9] amounting to about one-fifth of the total subsidies (explicitly) supplied to various industries. It is difficult to say how these implicit subsidies (item 3 above) are spent, however. There are three possibilities. They can be spent inefficiently on gorgeous office buildings, or transferred to bank clienteles in the form of low-interest-rate loans, or reinvested into business loans as a part of the retained earnings of banks, pro-

Table 4. Estimates of implicit subsidies and taxation on city banks, and total subsidies to private industries (annual average) (billion yen)

	1966–70	1971–75
1. Estimated implicit subsidies to city banks due to regulation of official discount rates	20	19
2. Estimated implict taxation to city banks due to regulation of bond yields	17	31
3. Estimated implicit subsides on city banks due to regulation of deposit rates	85	242
Total industrial subsidies	451	1,106

Sources: J. Teranishi (note 1), tables 8–14, 8–15, and 8–19; S. Ogura and N. Yoshino (note 13), table 4–1.

viding an increased capital base for intermediation. In view of the market-determined nature of bank loan rates, the possibility of the second happening is small. Since the dividend policy of banks was strictly regulated (in the form of a maximum dividend ratio) during the period, the third case cannot be neglected.[10]

Role of Government Finance

Throughout the period of high economic growth, there was a clear-cut complementary relationship between the private financial intermediaries and the government finance system.

On the liability side, there was division of labour in terms of the maturity period of deposits. The rates of interest on postal savings were kept higher for long-term deposits (more than two years) and lower for short-term deposits (6 to 18 months) than the deposit rate of private banks.[11] As a consequence, the ratio of average balance to total repayment during one year was much higher for postal savings than for private bank deposits; in 1965, 3.84 for the former and 0.85 for the latter.

On the asset side, the major part of the funds of private financial intermediaries was directed towards the growing or modern sector, especially that producing investment goods and exportables. This was more or less the consequence of the free play of market forces. Governmental guidance or intervention has never been truly effective.[12] On the other hand, most of the funds of government finance were supplied either to the declining traditional sector or to social overhead investment. Table 5, for example, shows that the shares of funds earmarked for the modernization of the low-productivity sector (agriculture and small and medium-sized firms) and providing overhead capital for

Table 5. Composition of fiscal investments and loans programme, by purpose (percentages)

	1956–60	1961–65	1966–70	1971–75
Promotion of key industries	16.6	9.9	6.3	3.7
Foreign economic aid and export promotion	4.3	7.9	10.4	8.8
Regional development	9.0	7.5	4.6	3.7
Overhead capital for industries	21.6	26.1	24.3	23.2
Modernization of low productivity sectors (agriculture and small- and medium-size firms)	20.9	19.0	20.1	19.6
Improvement of living conditions	27.6	29.6	34.3	41.0
Total	100.0	100.0	100.0	100.0

Source: S. Ogura and N. Yoshino (note 13).

industries (mainly railway and electricity) were significant, and that the share set aside for the promotion of key industries was low and decreasing. Similarly, according to table 3, smaller firms were more dependent on government finance in the 1957 manufacturing sector. The emphasis on the declining sector is more apparent in table 6. Such stagnating industries as mining, textiles, agriculture, and marine transportation were heavily dependent upon government finances for their investment activities, while such growing industries as steel, machines, and chemicals were not.[13]

To sum up, the pattern of the division of labour was as follows. The growing, modern sector obtained ample funds from private financial intermediaries at market interest rates, while funds for the declining, traditional sector and for social overhead provisions came from the government at regulated interest rates.

Three additional comments are in order. First, although the share of government finance was rather small, as revealed by table 1, the figures were higher in terms of shares in investment funds (excluding the working capital). Moreover, the ratio of postal savings to total private time and savings deposits steadily increased from 15 per cent in the 1950s to 40 per cent in the mid-1970s (levelling off thereafter). Second, although our analysis above is based on quantitative data such as tables 5 and 6, Sakakibara et al. regard the qualitative effects of government funds allocation as important; they argue that the government financial system actively sought after strategic industries where both expected profits and risk were high, fulfilling the role of a catalyst by concentrating large

Table 6. Percentage composition of government fund in total fixed investment, by industry

	1954–60	1961–67
Mining (coal mining)	25.7 (37.2)	39.9 (65.9)
Steel	4.6	3.6
Machine	11.3	9.5
Chemical	8.1	7.1
Textile	14.2	14.7
Agriculture and fisheries	52.9	47.9
Electricity	32.4	19.7
Marine transportation	33.9	50.9
Land transportation	10.4	21.9

Source: A. Horiuchi and M. Otaki (note 3).

amounts of funds in specific areas.[14] They conclude: "Had it not been for public financial intermediation, basic industries such as energy, steel, shipping, and petrochemicals would not have developed so smoothly." However, except for petrochemicals, none of the three other industries seem to belong to the high-return, high-risk category. Moreover, it must be recalled that the three industries have also benefitted significantly from government finance and from the rationing system of bond issuance in a quantitative sense. The importance of the qualitative effects is still open to question.

Finally, it must be added that bond issuance was mainly rationed to industrial overhead, so that it was complementary with government finance in its role of fund allocation. Electricity producers were the first, and by far the largest, issuers of corporate bonds, providing for 43 per cent of total issuance in 1956–1974. The steel sector was the second, its average share staying at about the 10 per cent level during the period.

Accommodating Stance in Monetary Policy

Since the balance of payments was more or less balanced in the medium term, and the open security market was not developed sufficiently, base money during the rapid growth period was mainly supplied through the Bank of Japan direct credit to (city) banks.

What is particularly interesting in this respect is that the Bank of Japan adopted a quite accommodating stance with regard to its credit supply, increasing credit whenever demand for it increased. As a result of this, its credit supply moved in the same direction as the call money rate. This feature was emphasized in Teranishi[15] and tentatively confirmed by Horiuchi and Otaki,[16] who indicate one-way Granger causality from income (manufacturing production index) and the nominal interest rate (call rate) to currency supply for monthly data from July 1960 to December 1972. Moreover, the call rate has positive and significant effects on currency supply in their estimate of the reaction function (base money supply function) of the Bank of Japan, based on monthly data from February 1961 to December 1971. The reason why the Bank of Japan took this particular policy stance seems to lie in the illiquidity problem of commercial banks. For one thing, a part of their assets was devoted to holding illiquid bonds which could not be resold on the market without loss. For another, they supplied investment funds to growing sectors with their relatively short-term deposits. On this point we have already underlined the difference in maturity period with postal saving. It must be noted also that the maturity period of deposits has considerably shortened because of the lowering of the wealth/income ratio in the early part of the rapid growth era. The share of time and saving deposits in total bank deposits was 50.4 per cent in 1935, but fell to 14.5 per cent in 1945, and only after 1960 returned to the pre-war level. Therefore, it can be conjectured that the accommodating stance of the Bank of Japan's credit supply was needed to alleviate these illiquidity problems.

Such a stance is liable to amplify the swings in the business cycle, however. It is worth noting that the Bank of Japan has utilized another very direct measure of restrictive policy in combination with this basic stance: this is the so-called "window guidance," which sets a ceiling on the quarterly rate of increase in city bank loans. It can be shown that window guidance and the accommodating credit supply are complementary in the sense that a necessary condition for the effectiveness of window guidance is the accommodating stance of Bank of Japan credit. In order to demonstrate this proposition, let us add call loan market to the asset-market model mentioned above, and divide the banking system into city and other banks. In the interbank call market, city banks behave as borrowers and other banks as lenders. The portfolio of private asset holders comprises base money, deposits, and other assets, as before. The rate of interest rate on other assets (i) represents the cost of capital of the economy: a decrease in i has expansionary effects on economic activity. In the deposit market, the interest rate is assumed to be fixed and equilibrium to be attained on the demand function of deposits by asset holders. It is also assumed that the deposits of city banks are a fraction (β) of total deposits, while the remaining fraction ($1 - \beta$) is held with other banks. The coefficient β is a constant, presumably dependent on the branch office licensing policy. For the sake of simplicity, let us delete α, W, and $\bar{R}$ from the demand for asset equations. Then, the demand for deposits and base money by private asset holders is given by C (i) and D (i) respectively, and dc/di < o, dD/di < 0. Let us finally assume that the base money is supplied solely through Bank of Japan credits to city banks.

Denoting the call rate and the Bank of Japan credits by ρ and N, respectively, the demand for call loan by city banks can be written as

$$\bar{A} + \gamma\beta D(i) - N - \beta D(i),$$

where γ is the fixed coefficient of bank reserve holding and $\bar{A}$, a constant, represents the constrained demand for other assets by city banks regulated by window guidance. (Here, window guidance is assumed to be related to stock, instead of the flow of loans, for the sake of convenience.) N takes the following values depending on the policy stance:

$$N = \frac{N(\rho)}{N} \quad (dN/dp > 0) \ldots\ldots \text{accommodating}$$
$$\phantom{N = \frac{N(\rho)}{N}} \quad (\text{const.}) \ldots\ldots\ldots \text{otherwise.}$$

The supply of call loan by other banks is represented as

$$(1-\beta)D(i) - A(i,\rho) - \gamma(1-\beta)D(i),$$

where $A(i,\rho)$ is demand for other assets by other banks, and $\partial A/\partial i > 0$, $\partial A/\partial \rho < 0$. (For the sake of simplicity, we have used the same coefficient of reserve holding as city banks.)

Equating the supply and demand of call loan, we obtain

$$\bar{A} + A(i,\rho) = N + (1-\gamma)D(i).$$

On the other hand, the equilibrium condition in the base money market is given by

$$N = \gamma D(i) + C(i).$$

These two equations represent a general equilibrium of assets markets for our model. The equilibrium condition for other assets can be deleted in the light of Walras' Law.

First, let us show that window guidance is ineffective in the absence of the accommodating stance. If $N = \bar{N}$ in (2), it can be readily confirmed that

$$di/d\bar{A} = 0,$$
$$d\rho/d\bar{A} > 0,$$
$$dA/d\bar{A} = -1.$$

A reduction of $\bar{A}$ due to window guidance is exactly offset by an increase in A, and has no effect on the cost of capital i. This is because a reduction of $\bar{A}$ decreases the demand for call loans by city banks, and the consequent fall of the call rate induces other banks to shift their asset holdings from call loans to other assets. This proposition was proved by Horiuchi.[17] Next, the effectiveness of window guidance, given the accommodating stance of the Bank of Japan's credit supply, can be readily demonstrated. If $N = N(\rho)$ in (2), we have:

$$di/d\bar{A} < 0,$$
$$d\rho/d\bar{A} > 0,$$
$$dA/d\bar{A} > -1.$$

With the accommodating stance, the fall in the call rate due to window guidance is smaller owing to the decrease in Bank of Japan credits. Consequently, the offsetting increase of the demand for other assets by other banks becomes smaller in absolute value than the reduction in $\bar{A}$. Therefore there occurs a net decrease in total demand for other assets $A + \bar{A}$, and the cost of capital is increased.

Now, a basic characteristic of monetary policy during the rapid-growth period can be summarized as follows. In order to alleviate the illiquidity of banks, Bank of Japan credit was supplied in a passive manner in accordance with the demand by banks. Window guidance was an effective tool of restrictive monetary policy under this accommodating stance of credit supply.

Two additional comments on the above analysis are in order. First, another necessary condition for the effectiveness of window guidance is the interest elasticity of the demand for reserves by banks. This can be readily confirmed by letting $\gamma = \gamma(\rho)$ in (1) and (2) with $N = \bar{N}$. Owing to the accommodating credit supply by the Bank of Japan and to the strong demand for funds by business firms, the reserve holding by banks tended to be kept at a minimum level. Therefore, the author is somewhat sceptical about the elasticity of reserve holding. However, this point must be settled empirically and is still open to question.[18] Second, it can be easily seen that the degree of effectiveness of window guidance is dependent on the magnitude of the partial derivative $\partial A/\partial i$, reflecting the degree of market segmentation of the loan market for other banks. The more window guidance is effective, the more segmented is the loan market. Although most other banks were required to specialize in certain types of loan clientele, this enforcement tended to become ineffective owing to the development of arbitrage techniques, such as lending through agent banks. Since this weakens the effectiveness of window guidance, the Bank of Japan placed some banks in the "other banks" category under window guidance, and the number of such other banks increased during the period.

International Capital Movements

The strict regulation of international capital transactions seems to have provided the necessary conditions for the financial system during the period, first in order to maintain proper lender–customer relationships and because they can produce information about borrowers jointly with the provision of other services, such as the provision of transaction accounts. Their accumulated informational capacity is most useful in coping with the massive emergence of new firms. It can be argued, therefore, that the high degree of financial intermediation during the high-growth period had growth-promoting effects through the efficient production of information.[24]

Although the theory outlined above is interesting and insightful, it is still open to the following question. A high degree of financial intermediation means a low degree of development of direct finance, especially of bond and equity markets. Apart from the problem of informational asymmetry, it has been shown that the equity market is more allocationally efficient than the loan market under general conditions of uncertainty.[25] Therefore, without a careful weighing of the informational advantages of intermediation with the allocational efficiency of the equity system, nothing definite can be said on this issue.

Adjustment Funds

Since the high growth involved drastic changes in industrial structure, the problem of allocation of funds to various industries, especially to the declining sector, was important. Generally speaking, the market-oriented private financial system has the following two drawbacks in the adequate supply of funds to the declining sector. First, whenever there is a lack of malleability of production factors, the market rate of return of the declining sector during the adjustment period tends to be lower than otherwise. In this case, the supply of funds through the free market system, whether of a direct- or an indirect-finance type, is liable to be inadequate.[26] Secondly, since the building up of informational channels between a firm and a bank has the property of fixed investment, it is particularly difficult to establish in the case of declining industries. As a consequence, even if the short-term rate of return of some firms within such industries is high, it is possible that adequate information is often not conveyed to banks. This is another factor which explains the possible lack of supply of funds to declining sectors.

In the case of Japan during the high-growth era, government finance is seen to have played the role of supplier of funds to the declining sector. This role can be understood as a remedy for the two market failures cited above. In this sense, the characteristic division of labour between government and private financial systems can claim special merit for its growth-promoting effects.

This is not the end of the story, however. It is worth noting that government finance has its own shortcomings. Since the area of activity of each governmental institution is narrowly specified, and it is not easy to change the legal stipulation, there occurred significant cases of inefficiency, such as a prolonged period of adjustment or excessive investment in declining sectors, possibly due to the inefficient and excessive loan supply by these institutions. This problem became the focus of debate when several major institutions reported a considerable amount of unused funds in the early 1970s. The reorganization of the government finance system has become a serious policy problem since then.

Concluding Remarks

Our analysis of the relationship between economic growth and financial regulation has been inconclusive. The effect on the level of capital cost was ambiguous. The efficiency of allocation under conditions of uncertainty is still unclear on the theoretical level. Although the adjustment role of government finance was theoretically growth-promoting, we cannot ignore possible cases of waste or of the inefficient use of funds.

However, after the mid-1970s the Japanese economy moved from a period of high growth into one of stable growth. At the same time, the financial system underwent drastic changes.[27] On the domestic side, the deposit rate was more or less liberalized as far as large denomination deposits were concerned. Money and bond markets have developed considerably; there is active non-bank participation and remaining regulations are only nominal. Internationally, foreign direct investment into Japan had been liberalized by 1973, except for some special sectoral restrictions. Portfolio transactions, by both residents and non-residents, have become almost free of regulations, except for medium and long-term transactions in Euro-yen.

It is worth while noting that there are three important features in the background to these changes:
1. The accumulation of private financial assets has advanced considerably. This has widened the scope for portfolio management by asset holders.
2. The domestic rate of profit has significantly declined. This seems to be one of the basic reasons for Japan's recent capital account deficits.[28]
3. The rapid change in industrial structure seems to have almost come to an end. The share of agriculture in net domestic product fell as low as 5.4 per cent in 1975.
4. Instead of business firms, the government and foreign sector have become the major deficit unit in terms of sectoral I-S balance. Since the government or foreign sector debts are highly homogeneous compared to business debt, this seems to have greatly reduced the informational needs of the economy.

Notes

1. For the long-term financial development of Japan, see H.T. Patrick, "Postwar Financial Development in Historical Perspective," mimeo (Yale University, 1982);

and J. Teranishi, *Nihon no keizai hatten to kinyu* (Money, Banking, and Capital in Japanese Economic Development) (Iwanami, Tokyo, 1982).
2. R. McKinnon, "Financial Repression and the Liberalization Problem with Less-developed Countries," in S. Grassman and E. Lundberg, eds, *The World Economic Order – Past and Prospects* (Macmillan, London, 1981).
3. A Horiuchi, "The 'Low Interest Rate Policy' and Economic Growth in Postwar Japan," *The Developing Economies*, vol. 22, no. 4 (1984); and A. Horiuchi and M. Otaki, "Sengo Nihon no kinyuseisaku" (Monetary Policy in Postwar Japan), paper presented at the Zushi Conference, March–April 1985.
4. Since average reserve holding by banks was 10–15 per cent of deposits, it can be seen that the deposit rate is still lower than the call rate even if adjustment is made for reserve ratio: deposit rate $(1 - \text{reserve rate}) \times$ call rate.
5. Y. Suzuki, *Money and Banking in Contemporary Japan* (Yale University Press, New Haven, Conn., 1980).
6. Possible effects on saving rates are ignored in this argument.
7. Similar perverse cases in the presence of the curb market are pointed out in S. van Wijnberger, "Interest Rate Management in LDCs," *Journal of Monetary Economics*, vol. 12, no. 3 (1983); and A. Kosaka, "The High Interest Rate Policy under Financial Repression," *The Developing Economies*, vol. 22, no. 4 (1984).
8. E. Sakakibara, R. Feldmar, and Y. Harada, *The Japanese Financial System in Comparative Perspective*, a study prepared for the use of the Joint Economic Committee (Congress of the United States, Washington, D.C., 1980), p. 40.
9. Item 3 is probably overestimated, since it includes the possible value added owing to the information-producing activities of banks (a part of α in figure 2).
10. Although the deposit (postal savings) rate of government finance was also regulated, it is clear that the implicit subsidies in this case were largely transferred to the clientele of government finance, as the loan rates were also regulated.
11. At the level of effective rates. The rate of interest on postal savings is compounded every six months.
12. See A. Horiuchi and M. Otaki (note 3 above).
13. These points were originally suggested by T. Ouchi in *Nihon keizai-ron (A Study on the Japanese Economy)*. (Todai-Shuppan, Tokyo, 1962), and recently confirmed by A. Horiuchi and M. Otaki (note 3 above) and S. Ogura and N. Yoshino, "Zeisei to zaisei-toyushi" (The Tax System and Fiscal Investments and Loans Programme), in R. Komiya, M. Okuno, and K. Suzumura, eds., *Nihon no sangyo seisaku* (Todai-Shuppan, Tokyo, 1984).
14. E. Sakakibara, R. Feldman, and Y. Harada (note 8 above). A similar view, especially with respect to the role of the Japan Development Bank, has been expressed by M. Higano, "Kinyukikan no shinsa noryoku to kokyaku kankei" (The Screening Capacity of Financial Institutions and the Lender–Customer Relationship), *Kikan gendai keizai*, 45 (1981).
15. See note 1 above.
16. See note 3 above.
17. A. Horiuchi, *Nihon no kinyuseisaku* (Monetary Policy in Japan) (Toyokeizai-Shinposha, Tokyo, 1980).
18. J. Teranishi (note 1 above), pp. 600–603.
19. For example, J. Sachs, "Aspects of the Current Account Behavior of OECD Eco-

nomies," NBER Working Paper, no. 859, February 1982.
20. Recall the Korean experience in the mid-1960s, where liberalization of the interest rates without regulation of the import of foreign capital caused significant inflation.
21. The high rate of profit apparently reflects capital scarcity, but it does not mean the high cost of preventing capital inflow. This is because the rate of capital accumulation remained very high without reliance on foreign capital.
22. H.S. Houthakker and S.D. Magee, "Income and Price Elasticities in World Trade," *Review of Economics and Statistics*, vol. Ll, no. 2 (1969).
23. A. Horiuchi and M. Otaki (note 3 above) make a similar point mainly in terms of 2.
24. To this may be added the possible inefficiency of new issue market of equity due to principal–agent problems (J.E. Stigler, "Information and Economic Analysis: A Perspective," *Economic Journal* (suppl., 95 (1985)). These points are essentially indebted to a discussion with Kermit Schoenholtz. The basic idea is his and any possible errors are mine.
25. R. Tachi and K. Hamada, *Kinyu* (Monetary Theory) (Iwanami, Tokyo, 1972). The transfer of future income through the loan market is carried out by fixed amount independent of the state of nature, while the equity market makes it possible depending on the state of nature. Needless to say, equity is still inferior to the state contingent claim system because the transfer by means of equity has to be a fixed fraction of the income of each state.
26. This is simply part of the general problem of adjustment aid. Cf. M. Mussa, "Government Policy and Adjustment Process," in J. Bhagwati, ed., *Import Competition and Response* (University of Chicago Press, Chicago, Ill., 1982; and S. Sekiguchi and T. Horiuchi, "Boeki to choseienjo" (Foreign Trade and Adjustment Aid), in R. Komiya, M. Okuno, and K. Suzumura, eds., *Nihon no sangyo seisaku* (Todai-Shuppan, Tokyo, 1984).
27. For details of the domestic side see S. Royama, "The Japanese Financial System: Past, Present and Future," *Japanese Economic Studies*, vol. 12, no. 2 (1983/84); and Y. Suzuki, "Changes in Financial Asset Selection and the Development of Financial Markets in Japan," *Monetary and Economic Studies* (Bank of Japan), vol. 1, no. 2 (1983). For the international aspects, see Y. Shinkai, "Internationalization of Finance in Japan," unpublished (1984).
28. Other reasons are (a) the difference in the movements of the budget deficits of the US and Japan, and (b) the difference in the corporate tax system of the two countries.

FINANCIAL LIBERALIZATION AND THE INTERNAL STRUCTURE OF CAPITAL MARKETS: THE PHILIPPINE CASE

Mario B. Lamberte

Introduction

In the 1950s and 1960s, the Philippines launched various programmes to accelerate economic development. To stimulate investment in certain economic activities, especially in import-substituting industries, interest rates were administratively set at levels significantly lower than the prevailing market rates. Whenever the credit needs of certain priority sectors were not satisfied, specialized financial institutions, such as rural banks, development banks, etc., which mainly relied upon government and Central Bank support, were created. The whole financial policy framework then idealized investment and relegated savings mobilization to the background. The result was severe financial repression and fragmentation that failed to develop the capital market, especially the long-term capital market.

In the early part of the 1970s, the Philippines recognized the need to mobilize domestic savings and provide a base for the development of the long-term capital market. Financial reforms intended primarily to liberalize the financial system were introduced in the mid-1970s. This was followed by another set of financial reforms in the early 1980s.

This paper analyses the Philippines' experience with financial repression and liberalization. The specific objectives are: (1) to re-examine the traditional financial policies that repressed financial development; and (2) to analyse the objectives and results of financial liberalization efforts.

The views expressed in this paper are those of the author and do not necessarily reflect those of the Philippine Institute of Development Studies (PIDS).

In the following section, which is the main body of this study, the current capital market of the Philippines is briefly described. The section is divided into three subsections: the first deals with traditional financial policies; the second with the initial liberalization efforts; and the third with the most recent financial reforms. The last section concludes the study and also presents some suggestions for strengthening the capital market.

A Brief Description of the Capital Market[1]

This section describes the components of the capital market, namely the loans, bonds, and equity markets. The money market is also described because it serves as a catalyst for mobilizing funds.

The Loans Market and the Financial System

In terms of the variety of financial institutions and the array of financial instruments offered, it can be said that the Philippine financial system has already achieved a certain degree of sophistication compared with other Asian countries. Table 1 presents the assets and number of institutions in the financial system as of December 1983. The system is dominated by commercial banks whose assets comprised 56 per cent of the total assets in 1983. The presence of government in the financial system is highly visible. It owns the biggest commercial bank (PNB), the biggest development bank (DBP), and the biggest insurance agencies (GSIS and SSS). Apart from this, government-owned financial institutions also own or control a number of commercial banks that used to be controlled by the private sector. Specifically, DBP controls the Associated Bank; GSIS, the Commercial Bank of Manila; NDC, the International Corporate Bank; SSS and the Land Bank, the Union Bank; and PNB, the Pilipinas Bank. This is not to mention the equity exposure of DBP in several private development banks. Unlike specialized government financial institutions, these banks compete with private banks in both the deposit and loan markets. Because of their special relations with government, they have an undue advantage over privately owned commercial banks.

Figure 1 presents a simplified sketch of the flow of funds from primary savers to the ultimate users of funds, emphasizing the presence of government financial institutions. These tend to be large and are organized for specific purposes. Except for PNB, government-owned financial institutions do very little intermediation outside their captive depositors, relying more heavily instead on budgetary support, and Central Bank and foreign borrowings.

The figure recognizes the presence of offshore banking units (OBUs) in the financial system. Their task is to augment domestic resources, and they lend either directly to non-bank entities or indirectly through foreign currency deposit units (FCDUs).

Table 1. Philippine financial system in 1983: total assets, shares, and number of institutions

	Total assets (millions of pesos)	Shares (%)	Number of institutions
Central bank	130,371.7	—	1
Financial system	354,606.2	84.8	
Banking institutions	326,013.0	78.0	1,122
Commercial banks	235,040.0	56.2	34
PNB	70,502.3	16.9	1
Thrift banks	16,149.0	3.9	136
PNBs	4,613.2	1.1	45
Savings and mortgage banks	7,399.4	1.8	8
Stock SLAs	4,136.4	1.0	83
Rural banks	9,499.7	2.3	949
Specialized govt banks	65,323.9	15.6	3
DBP	56,529.7	13.5	1
Land bank	8,530.2	2.0	1
Philippine Awanah bank	264.0	0.1	1
Non-bank financial institutions	28,593.2	6.6	1,474
Investment houses	7,210.4	1.7	14
Finance companies	11,810.8	2.8	336
Investment companies	6,159.9	1.5	65
Securities dealers/brokers	683.0	0.2	124
Pawnshops	483.1	0.1	701
Fund managers	1,529.6	0.4	12
Lending investors	49.2	0.0	120
Non-stock SLAs	648.3	0.2	74
MBLA	18.9	0.0	7
Other financial institutions	63,485.2	15.2	
Private insurance companies	13,715.7	3.3	136
Specialized non-banks	49,658.3	11.9	
GSIS	14,707.2	3.5	1
SSS	16,227.1	3.9	1
ACA	—	—	
NIDC, PHIVIDEC, NHMFC and NDC	18,724.0	4.5	
Venture Capital Corp.	111.2	0.0	15
Offshore banking units (OBUs)	4,408.0[a]	—	21
Total (excluding Central Bank and OBUs)	418,091.4	100.0	

a. In US dollars.
Source: Nomura Research Institute (1984), p. 16, except line D.

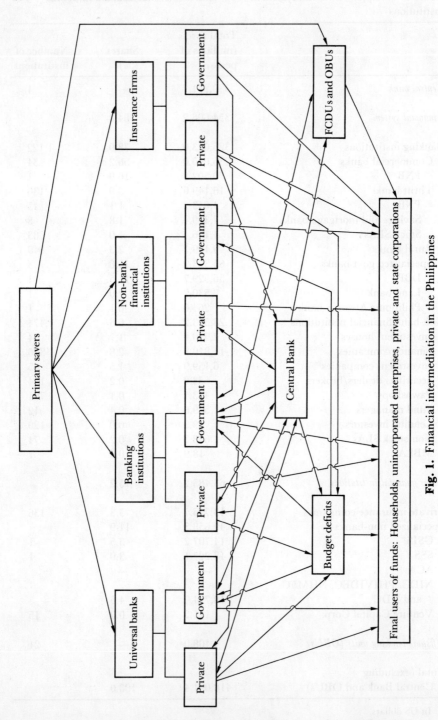

Fig. 1. Financial intermediation in the Philippines

The figure also shows the flows between institutions. For example, the two government insurance agencies are required by law to invest a portion of their funds in DBP financial instruments, which, in turn, are used for long-term lending. The reserve requirements of banks and non-banks authorized to perform quasi-banking (NBQBs) lead the flow of funds into the Central Bank, which, in turn, rechannels the financial intermediation tax to specialized government banks and/or private banks for specific purposes.

There is one aspect of figure 1 that is also worth noting, that is, the role of government budget deficits. These deficits are used to support government financial and non-financial institutions. If the fiscal sector decides to finance its deficits by borrowing from the Central Bank and deposit money banks, then less funds will flow to private productive economic agents. It should be noted that the Finance Minister usually finds it easier to raise revenue through inflation tax rather than through raising ordinary taxes to finance certain activities, since he is also a member of the Monetary Board. Indeed, recent experience shows that in times of great need the Finance Minister will resort to such a practice.

Commercial banks have long dominated the loan market. Of the total loans outstanding in 1983, 60 per cent belonged to commercial banks (see table 2). A big chunk of the remaining portion was provided by DBP and government insurance agencies. It is to be noted that commercial banks' loans were concentrated in the shorter end. Only about 31 per cent of the total loans outstanding in 1983 were intermediate- and long-term.

Table 2. The Philippine financial system: total loans outstanding

	1983 (PM)	Percentage of total
Commercial banks	108,643.0	59.8
PNB	37,123.5	28.4
Development banks	39,094.1	21.5
DBP	30,919.7	17.0
Savings and mortgage banks	4,765.8	2.6
Rural banks	7,648.0	4.2
Private NBFI	2,278.3	1.2
Investment homes	1,770.3	1.0
Thrift NBFI	508.0	0.2
Private insurance companies		
Government NBFI	19,341.5	10.6
GSIS + SSS	17,080.8	9.4
NIDC	2,260.7	1.2
Total	181,770.7	100.0

Source: Central Bank of the Philippines.

The Bond Market

The bond market is almost synonymous with the market for government securities. Private bond issues have been very rare.

The government securities used in the market are presented in table 3. National government issues dominate government securities, followed by Central Bank and government corporate issues, in that order. Treasury Bills and CB Bills, which are fairly recent issues, are the only short-term securities.

Most government securities carry fixed interest rates far below the market rates. However, to make them marketable, some "sweeteners" are added; for example, they could be eligible as reserves against deposits (see Appendix). Indeed, regulations have provided them with a captive market. Of the total outstanding government securities in 1983, 36 per cent were held by commercial banks, and 27 per cent by government financial entities. Only 10 per cent were in the hands of the private sector.

A secondary market for government securities hardly exists, except for Treasury Bills and CBCIs. The latter are actively traded as money-market instruments.

The Equity Market

The equity market has been an unimportant source of long-term finance. Primary issues have been thin. Likewise, the stock market has remained fairly narrow. Only 58 companies belonging to the top 1,000 corporations are listed in the two stock exchanges. The stock market is considered by many as a gambling casino where only speculative stocks, such as mining, are actively traded. It has been on the downtrend since 1978. Total volume traded in 1983 amounted to only 37,113 million compared with 151,363 million shares in 1978.

The Money Market[2]

The money market is a fairly recent phenomenon. It started in the mid-1960s when a few investment houses began buying and selling short-term dated debt instruments of banks and prime corporate names. Shortly thereafter, some commercial banks joined the bandwagon.

The money market was very active, especially in the 1970s and early 1980s. Since 1983, however, it has been on the downtrend owing to the collapse of a number of financial firms and the generally depressed economy. The money market is basically a wholesale market where lots of more than 50,000 pesos are densely traded. However, roughly 90 per cent of the total transactions have maturities of 45 days or less.

Interbank call loans are the primary money-market instruments, followed by promissory notes (see table 4). Commercial banks are the primary investors in this market, followed by other private corporations (see table 5).

Table 3. Outstanding government securities issued by and through the Central Bank (thousands of pesos, as of 31 May 1984)

Type of security	Outstanding	Percentage of total
National govt issues	33,110,585	74.4
PWED bonds	516,236	1.2
Premyo savings bonds	1,401,536	3.1
NPC capital bonds	321,600	0.7
NPC bonds	336,579	0.8
MWSS capital bonds	155,000	0.3
NLC	8,345	0.0
NIA bonds	49,427	0.1
Treasury bonds	12,604,187	28.3
Capital treasury bonds	400,000	0.9
Capital bonds, non-interest-bearing	200,000	0.4
Treasury capital bonds	1,135,300	2.5
Capital treasury notes	415,000	0.9
Treasury notes	10,215,893	22.9
Long-term	1,651,073	3.7
Medium-term	8,564,821	19.2
Treasury bills		
Short-term	5,351,481	12.0
Govt corporate issues	3,998,427	9.0
NPC bonds	180,700	0.4
NAWASA bonds	32,739	0.1
CCP bonds	139,640	0.3
EPZA bonds	433,000	1.0
Philippine charity sweepstakes Office bonds	240,644	0.5
Public estates	440,000	1.0
NDC bonds	600,000	1.3
NHA bonds	178,175	0.4
NFA bonds	800,000	1.8
Bahayan certificates	668,530	1.5
LRTA notes	285,000	0.6
Central Bank issues	7,419,370	16.7
CBCIs	5,421,870	12.2
Central Bank bills	1,997,500	4.5
Total	44,528,382	

Source: Nomura Research Institute (1984).

Table 4. Volume of money-market transactions by type of instruments, 1982

	PM	Percentage of total
Inter-bank call loans	20,694.8	44.2
Promissory notes	18,860.6	40.2
Repurchase agreements (private)	680.3	1.4
Repurchase agreements (government)	2,225.0	4.7
Certificate of assignments (private)	12.6	0.0
Certificate of assignments (government)	—	—
Certificate of participation (private)	6.8	0.0
Certificate of participation (government)	—	—
Commercial papers (non-financial)	1,055.7	2.2
Commercial papers (financial)	1,930.3	4.1
CBCIs	860.9	1.8
Treasury Bills	236.8	0.5
DBP bonds	163.0	0.3
Other government securities	136.0	0.3
Total	46,862.8	100.0

Source: Philippine financial statistics, 1982.

Table 5. Volume of money-market transactions classified by type of investor, 1982

	PM	Percentage of total
Individuals	2,973.0	6.3
Commercial banks	23,753.8	50.7
Rural thrift banks	1,148.7	2.4
Other banking institutions	3,385.4	7.2
Investment houses	4,388.2	9.4
Investment companies	327.7	0.7
Finance companies	2,296.6	4.9
Trust/pension funds	1,216.1	2.6
Government insurance	—	
Private insurance companies	620.1	1.3
Other government corporations	600.4	1.7
Other private corporations	5,792.3	12.4
Securities dealers	121.3	0.2
Lending investors	35.9	0.1
National government	3.3	0.0
Local governments	—	
Total	46,862.8	100.0

Source: Philippine financial statistics.

Financial Liberalization Efforts: Objectives and Results

This section discusses the Philippines' experience with financial repression and liberalization during the period 1956–1984. Following Laya,[3] the whole period is divided into three subperiods according to the kind of interest-rate policy adopted by the Monetary Board. These are: (1) rigid and low interest-rate policy period, 1956–1973; (2) transition period, 1974–1980; and (3) floating interest-rate period, 1981–1984.

Rigid and Low Interest-rate Period, 1956–1973

The years from 1956 to 1973 are considered to be the period of rigid financial repression. This can be gleaned from the various financial measures instituted during that period.

After the Second World War, a fairly modern banking system started to emerge, mainly dominated by five big branches of foreign banks and two government-owned banks. However, banks were primarily engaged in extending merchant drafts and in supplying credits to their clients. Only one government-owned bank was really engaged in long-term financing.

In contrast, the bond and equity markets were grossly underdeveloped. Private and government securities were practically nil. The Manila Stock Exchange, which was established in 1927, was performing unsatisfactorily with a few stocks listed and being traded. It was against this background that the Central Bank was established.

The Central Bank of the Philippines started operating in 1949, and was designed to be a development-oriented central bank. Its immediate task was to develop the financial structure of the economy. Thus, aside from the usual stabilization function of an ordinary central bank, the Central Bank was also given the responsibility of promoting a rising level of production, employment, and real income. In order to achieve this it allocated credit to priority sectors through its rediscount windows and also provided credit to government financial institutions, or to specific financial institutions like the rural banks. A high degree of co-ordination between fiscal and monetary policies was assured by making the Finance Minister a member of the Monetary Board. Central Bank regulatory controls extended to all banking institutions, not only to deposit money banks. It was also given the responsibility of creating specialized financial institutions and developing a government securities market.

The authorities responded fairly quickly to the unfilled credit needs of certain sectors by creating specialized financial institutions. Thus, rural banks were created in 1952 to service the credit needs of farmers in the countryside; savings and loan associations were established in 1963, primarily to service the needs of households by providing personal finance and long-term financing for home building and development; private development banks were organized in 1964

Table 6a. Structure of interest rates: commercial banks (per cent per annum)

Date	CB circular no.	Savings deposits	Time deposits					
			90	180	270	360	540	730
25 May 1956	67	2		2		2.5		
2 Sept. 1957 (effective Oct. 1957)	74	3	2	3		3.25	3.5	
26 Oct. 1960	112	3	Maximum interest rate on time deposits shall be 4% p.a.					
27 March 1963	149	3.5	3.75	4		4.25	4.5	
15 Dec. 1964	185	4	4.25	4.5		4.75	5	
20 Jan. 1965	190	4.5						
10 Nov. 1965	212	5.75	5.75	6		6.25	6.5	
6 June 1967	239	5.75	5.75			6		
16 Apr. 1969	272	6		6.5		7		
21 Feb. 1970	292	6		6.5		7		8

Source: Various CB circulars.

Table 6b. Structure of interest rates: thrift banks and rural banks (per cent per annum)

Date	CB circular no.	Savings deposits	Time deposits					
			90	180	270	360	540	730
2 Sept. 1957 (effective Oct. 1957)	74	3	2	3		3.25	3.5	
25 Oct. 1957	78	3.5						
26 Oct. 1960	112	3.5	Maximum interest rate on time deposits shall be 4% p.a.					
27 March 1963	149	4	3.75	4		4.25	4.5	
9 Dec. 1963	161	4		4.25		4.5		
15 Dec. 1964	185	4.5		4.75		5		
10 Nov. 1965	212	5.75	5.75	6		6.25	6.5	
16 Apr. 1969	272	6		6.5		7		
21 Feb. 1970	292	6		6.5		7		8

Source: Various CB circulars.

to provide medium- and long-term loans. However, rigid Central Bank regulations forced them to perform highly specialized functions. This inevitably contributed to a high degree of fragmentation in the financial system.

The Central Bank then did not have the power to set lending rates. The latter were governed by the Usury Act of 1916 which prescribed ceilings of 12 and 14 per cent for secured and unsecured loans respectively.

The interest ceilings on loans obliged the Central Bank to prescribe interest ceilings on deposits starting in 1956. The schedules of those ceilings, shown in tables 6a and 6b, are clearly suggestive of the Central Bank's policy. First, the interest-rate differentials between commercial banks on the one hand and thrift and rural banks on the other suggest that the latter were given preferential treatment since they were most likely to provide medium- and long-term loans and to support the government's programme. However, the interest-rate differentials were hardly noticeable. Second, the interest-rate differential between savings and time deposits was set to encourage savers to place their funds in long-term financial assets. Again, the differential between these two types of deposits was very minimal. Third, the low interest ceilings allowed banks to enjoy a substantial profit margin of between 9 and 10 percentage points. It could be that the wide profit margin allowed by the Central Bank was meant to encourage more entrants into the financial system – indeed, the Central Bank was very lax in giving licences to potential entrants. Finally, the interest rates on deposits were adjusted upwards, but at long time intervals and in smaller steps.

In the 1950s, the Central Bank was very much preoccupied with defending the exchange rate, which was fixed at 2 pesos to the dollar, by foreign exchange and import controls. The strategy was to allow only a modest growth in money supply in order to control inflation. In the 1960s, however, the Bank started to promote economic development intensively by the active use of its selective credit control. It increased its basic rediscount rate and, at the same time, began to give preferential or concessional rediscount rates to certain priority sectors (see table 7). It is to be noted that only short-term notes were eligible for rediscounting.

The fairly wide margin between the discount and prescribed loan rates tended to make the demand for borrowing from the Central Bank infinitely elastic. In this case, "the level borrowed was determined by what the Central Bank decided to lend to the banking system."[4] This was the beginning of the banking system's dependence on the Central Bank for cheap funds.

During this period, the Central Bank took the opportunity of increasing the tax on financial intermediation. Reserve requirements on the savings and time deposits of commercial banks were gradually raised from 5 per cent in 1959 to 20 per cent in 1970 (see table 8). The movements of the reserve requirement ratios, especially in the early 1970s, seem to suggest that this monetary instrument was used mainly to exact a tax on financial intermediation, not to control money supply movements. This is indeed another dimension of financial repression during the period.

The Central Bank gave substantial support to bond issues by the govern-

Table 7. Schedule of changes in Central Bank rediscount rates to commercial banks (1949–1973)

Circular, MAAB, or MB Resolution		Basic rate (1)	Agricultural crop loans (2)	Preferential rate				Rice and corn financing (7)
No.	Date			Agricultural packing credits and export bills (3)	Agricultural papers (4)	Industrial papers (5)	Export financing (6)	
MB Res. 164	1949: 4 Aug.	1.50	—	—	—	—	—	—
MB Res. 317	29 Dec.	3.00	—	—	—	—	—	—
MB Res. 491	1952: 7 Aug.	2.00	—	—	—	—	—	—
MB Res. 75	1954: 19 Jan.	1.50	—	—	—	—	—	—
MB Res. 378	1957: 1 Apr.	2.00	—	—	—	—	—	—
MB Res. 1226	28 Aug.	4.50	—	—	—	—	—	—
MAAB	1959: 3 Feb.	6.50	4.50	5.00	—	—	—	4.50
MAAB	1960: 7 June	6.00	4.50	5.00	—	—	—	4.50
MAAB	8 Sept.	5.75	4.50	5.00	4.50	4.50	—	4.50
MAAB	21 Nov.	5.00	4.00	4.00	4.00	4.00	—	4.00
MB Res. 719	1961: 5 May	3.00a	3.00	3.00	3.00	3.00	—	3.00
MAAB	1962: 8 Jan.	6.00	6.00	6.00	6.00	6.00	—	6.00
MB Res. 985	21 Aug.	6.00	3.00	3.00	3.00	3.00	—	3.00
MB Res. 121	1966: 18 Jan.	4.75	3.00	3.00	3.00	3.00	—	3.00
Cir. No. 242	1967: 23 June	6.00	2.00	3.00	2.00	2.00	4.75	3.00
Cir. No. 256	1968: 27 Feb.	7.50	2.00	3.00	2.00	2.00	5.75	4.00
MAAB	1969: 16 Apr.	8.00	2.00	3.00	2.00	2.00	5.75	4.00
Cir. No. 276	1969: 17 June	8.00	2.00	3.00	2.00	2.00	5.75a	4.00b
Cir. No. 291	1970	10.00	2.00	3.00	2.00	2.00	7.75	6.00

a. 6 per cent for textiles and other similar industries listed in the most recent Export Priorities Plan of the BDI pursuant to MAAB dated 16 May 1973.
b. May be availed of at 3 per cent Masagana 99.

Source: Department of Economic Research, Central Bank.

Table 8. Legal reserve requirements against peso deposit liabilities of commercial banks (1949–1981)

Effective date	Reserve ratio (per cent)			
	Demand	Savings	Time	"Now" accounts
10 Jan. 1949	18	5	5	
2 Feb. 1959	19	5	5	
4 Mar. 1959	20	5	5	
3 Apr. 1959	21	5	5	
7 Sept. 1960	19	5	5	
21 Nov. 1960	18	5	5	
21 Dec. 1961	17	5	5	
21 Jan. 1961	16	5	5	
15 May 1961	15	5	5	
21 Jan. 1962	19	5	5	
5 Aug. 1963	19	6	6	
19 Mar. 1965	10	10	10	
24 Jan. 1966	10	8	5 and 6[a]	
26 June 1967	11	8	8	
26 July 1967	12	8.5	8.5	
26 Aug. 1967	13	9	9	
26 Sept. 1967	14	10	10	
31 Oct. 1967	15	10	10	
30 Nov. 1967	16	12	12	
31 Dec. 1967	16	14	14	
31 Jan. 1968	16	16	16	
16 Feb. 1970	17	17	17	
2 Mar. 1970	18	18	18	
1 May 1970	18.5	18.5	18.5	
1 June 1970	19	19	19	
1 July 1970	19.5	19.5	19.5	
1 Aug. 1970	20	20	20	
30 June 1975	20	20	20[b]	
17 Jan. 1977[c]	20	20	20[b]	
26 Sept. 1979[d]	20	20	20[b]	
22 Aug. 1980	20[e]	20[e]	20[e]	
27 Feb. 1981	16[f]	16[f]	16.5[f]	16[f]

a. 5 per cent on liabilities of more than 30 days maturity; 6 per cent on liabilities with maturities of up to 30 days.
b. Time deposits with remaining maturities of more than 730 days are exempt from reserve requirements.
c. CBCI holdings of commercial banks used as reserves should not exceed 10 per cent of maximum required reserved securities and/or cash in vaults.
d. Land Bank and Awanah Bank are considered commercial banks insofar as reserve requirements are concerned.
e. Universal banking was approved under Cir. No. 752, dated 23 August 1980, to maintain a 20 per cent reserve requirement on all types of deposit liabilities.
f. Under CB Cir. No. 782, dated 27 February 1981, the reserve requirements against demand, savings, "now" accounts and time deposits with original maturity of 730 days or less shall be 19 per cent and shall be decreased at the rate of one percentage point every semester thereafter until the reserve requirement shall have been decreased to 16 per cent, while time deposits with original maturity of more than 730 days shall be 1 per cent and shall be increased at the rate of one percentage point every semester thereafter until the 5 per cent requirement shall have been reached.

Source: CB Statistical Bulletin, 1982.

ment. However, the yields on government securities were less attractive, especially, in the 1960s when rates of competing financial assets started to rise. Moreover, the prices of government securities, except the 7 per cent government bond, were fixed in at par in the past. This means that the original value (par value) of the government securities were kept constant, and it was also at that value that they were redeemed. The failure to adjust the features of government securities was due to the fact that the government then "considered only the interest cost or debt servicing of the issues when deciding on the rates and not their potential contribution to financial development or to banking stability."[5] In order not to lose entirely the market for government securities, the Central Bank decided in the mid-1960s to make the securities eligible as bank reserves.

We have outlined above the financial policies of the period 1956–1973, which in general tended to be repressive. We will discuss below the consequences of those policies.

Table 9 presents information regarding the flow of loanable funds and other pertinent economic indicators. McKinnon's measures of the flow of loanable funds are used in this study since, as observed above, the banking system still dominates the financial system. For this section, only the data for the period 1956–1973 will be discussed. The table shows a rising M2/M1 ratio, indicating the growing preference of wealth holders for higher-yielding financial assets.[6] The M2/GNP ratio started at 19.4 per cent in 1956, slowly rose to 26 per cent in 1963, then dropped to 23.4 per cent in 1965. It started to go up again in 1966 and stayed at relatively higher levels up to 1969. Note that during this period real interest rates on deposits were positive. The ratio started to decline in 1970, reaching 19.4 per cent in 1973. The severely negative real interest rates on deposits during this time must have caused disintermediation as wealth holders shifted to assets less vulnerable to inflation. If deposit substitutes are included, however, the M3/GNP ratio shows a higher level in 1973, partly confirming our contention that wealth holders shifted to high-yielding, short-term financial assets during a severe inflationary period.

It is worth while to compare the Philippines' performance during the period 1956–1973 with that of the Republic of Korea (see table 10). In 1956, the Philippines' M3/GNP ratio was almost twice as high as Korea's. In 1973, however, while the Philippines' M3/GNP ratio reached 25 per cent, Korea's increased almost fourfold. It is noteworthy that between 1963 and 1965 Korea undertook interest reforms, while the Philippines maintained its rigid and low interest-rate policy for the entire period from 1956 to 1973. Thus it can be said that interest-rate repression in the Philippines impeded the development of an organized financial system.

There is clear evidence that the actual lending rates were below the statutory rates up until the mid-1960s (see table 11). This was mainly due to the low demand for funds at the prevailing rates. Thus the prescribed lending rates were not distortionary before the mid-1960s. Although the situation had

Table 9. Flow of loanable funds, inflation rate, nominal and real interest rates[a]

					Nominal interest rates				
Regimes	M2/M1[b,c]	M3/M2[d]	M2/GNP	M3/GNP	Savings deposits	Time deposits (360 days)	WAIR[e]	Short-term loans (secured)	Long-term loans (secured)
Low interest-rate period, 1956–1973									
1956	1.36	1.00	19.4	19.4	2	2.5	—	12.0	12.0
1957	1.39	1.00	19.4	19.4	3	3.5	—	12.0	12.0
1958	1.41	1.00	20.4	20.4	3	3.5	—	12.0	12.0
1959	1.45	1.00	20.5	20.5	3	3.5	—	12.0	12.0
1960	1.51	1.00	20.4	20.4	3	4	—	12.0	12.0
1961	1.64	1.00	23.5	23.5	3	4	—	12.0	12.0
1962	1.78	1.00	25.1	25.1	3	4	—	12.0	12.0
1963	1.82	1.00	26.0	26.0	3.5	4.5	—	12.0	12.0
1964	1.85	1.00	24.1	24.1	4	5	—	12.0	12.0
1965	1.84	1.00	23.4	23.4	5.75	6.5	—	12.0	12.0
1966	1.98	1.00	24.8	24.8	5.75	6.5	—	12.0	12.0
1967	2.08	1.00	27.5	27.5	5.75	6	—	12.0	12.0
1968	2.08	1.00	25.7	25.7	5.75	6	—	12.0	12.0
1969	1.99	1.00	26.2	26.2	6	6	—	12.0	12.0
1970	2.00	1.00	23.0	23.0	6	7	—	12.0	12.0
1971	2.03	1.00	12.2	21.2	6	7	—	12.0	12.0
1972	1.83	1.00	21.2	21.2	6	7	—	12.0	12.0
1973	1.93	1.29	19.4	25.0	6	7	—	12.0	12.0
Average	1.78	1.02	22.8	23.2					

Table 9 *(continued)*

Regimes	Savings deposits	Time deposits (360 days)	WAIR	Real interest rates Short-term loans (secured)	Long-term loans (secured)	CPI	GNP
Low interest-rate period, 1956–1973							
1956	(0.8)		—			2.80	5.28
1957	1.4	(0.3)	—			1.60	3.97
1958	(0.13)	1.9	—	9.20	9.20	3.13	6.25
1959	4.87	0.37	—	10.40	10.40	(1.87)	1.36
1960	(2.00)	5.37	—	8.87	8.87	5.00	6.94
1961	(1.31)	(1.00)	—	13.87	13.87	4.31	5.51
1962	(0.26)	(0.31)	—	7.00	7.00	3.26	6.95
1963	(4.71)	0.74	—	7.69	7.69	8.21	3.42
1964	(4.75)	(3.71)	—	8.74	8.74	8.75	5.03
1965	2.53	(3.75)	—	3.79	3.79	3.22	4.35
1966	1.07	3.28	—	3.25	3.25	4.68	4.79
1967	0.12	1.82	—	8.78	8.78	5.63	5.56
1968	3.56	0.37	—	7.32	7.32	2.19	5.16
1969	4.77	3.81	—	6.37	6.37	1.23	3.96
1970	(8.85)	5.77	—	9.81	9.81	14.85	6.43
1971	(15.90)	(7.85)	—	10.77	10.77	21.90	5.39
1972	(2.22)	(14.90)	—	(2.85)	(2.85)	8.22	9.36
1973	(10.50)	(1.22)	—	(9.90)	(9.90)	16.50	5.28
		(9.50)		3.78	3.78		
				(4.50)	(4.50)		
Average	(1.84)	(1.06)	—	5.69	5.68	6.31	1.79

FINANCIAL LIBERALIZATION: THE PHILIPPINE CASE 217

Regimes	Nominal interest rates								
	M2/M1	M3/M2	MS/GNP	M3/GNP	Savings deposits	Time deposits (360 days)	WAIR	Short-term loans (secured)	Long-term loans (secured)
Transition period, 1974–1980									
1974	1.85	1.44	16.8	24.34	6	9.5	17.57	12.0	12.0
1975	1.27	1.50	16.8	25.2	6	9.5	15.01	12.0	12.0
1976	2.07	1.43	18.6	26.8	7	10	12.94	12.0	19.0
1977	2.18	1.35	21.2	28.7	7	10	12.59	12.0	19.0
1978	2.38	1.28	22.8	29.3	7	10	10.72	12.0	19.0
1979	2.41	1.26	20.8	26.3	9	12	12.89	14.0	21.0
1980	2.46	1.22	21.0	25.6	9	14	13.27	14.0	21.0
Average	2.18	1.35	19.7	26.6				6.14	

Table 9 (*continued*)

Regimes	Savings deposits	Time deposits (360 days)	WAIR	Real interest rates Short-term loans (secured)	Long-term loans (secured)	CPI	GNP
Transition period, 1974–1980							
1974	(28.16)	(24.66)	(16.59)	(22.16)	(22.16)	34.16	5.6
1975	(0.78)	2.72	8.23	5.22	5.22	6.78	6.01
1976	(2.23)	0.77	3.71	2.77	9.77	9.23	7.32
1977	(2.93)	0.07	2.66	2.07	9.07	9.93	6.37
1978	(0.28)	2.72	3.44	4.72	11.72	7.28	5.75
1979	(7.51)	(4.51)	(3.68)	(2.51)	4.49	16.51	6.91
1980	(8.60)	(3.60)	(4.33)	(3.60)	3.40	17.60	5.07
Average	7.50	4.07	(0.93)	−1.93	3.07	14.50	6.14

Regimes	M2/M1	M3/M2	MS/GNP	M3/GNP	Nominal interest rates Savings deposits	Time deposits (360 days)	WAIR	Short-term loans (secured)	Long-term loans (secured)
Floating interest-rate period, 1981–1984									
1981	2.67	1.31	20.7	27.0	9.79	15.60	15.60	16.00	21.08
1982	3.35	1.21	23.5	28.4	9.78	14.21	14.21	17.13	21.74
1983	2.95	1.18	25.3	29.8	9.69	14.34	16.60	21.28	23.41
1984	3.27	1.09	20.5	22.5	11.56	32.48	27.16	39.10	45.53
Average	3.06	1.20	22.5	26.9					

Regimes	Savings deposits	Time deposits (360 days)	WAIR	Real interest rates				
				Short-term loans (secured)	Long-term loans (secured)	CPI	GNP	
Floating interest-rate period, 1981–1984								
1981	(2.60)	3.21	3.21	3.61	8.69	12.39	3.45
1982	(0.44)	4.00	4.00	6.92	11.54	10.21	1.56
1983	(0.48)	4.17	6.43	11.11	13.24	10.17	1.26
1984	(38.79)	(17.87)	(23.19)	(11.25)	(4.82)	50.35	(5.59)
Average	(10.58)	(1.62)	(2.39)	2.60	7.16	20.73	0.19

a. Prior to 2 January 1976, all loans were subjected to interest-rate ceilings prescribed by the provisions of the Usury Law. Circular No. 783 dated 27 February 1981 lifted the ceilings on secured and unsecured loans of more than 730 days. Definitions of short-term and long-term were changed on 1 October 1981 from less than 730 days to less than 365 days and from more than 730 days to more than 365 days, respectively. Rates starting in 1982 are based on weighted average interest rates on loans of eight selected banks.
b. M1 = Currency in circulation + demand deposits.
c. M2 = M1 + savings and time deposits.
d. M3 = M2 + deposit substitutes.
e. WAIR = Weighted average interest rates of deposit substitutes.
Source: Department of Economic Research, Central Bank.

Table 10. A comparison of the flow of loanable funds: Republic of Korea v. Philippines

Year	Republic of Korea (M2/GNP)	Philippines (M2/GNP)
1956	10.8	19.4
1960	12.2	20.4
1970	34.7	23.0
1973	40.3	19.4

Sources: Republic of Korea, *International Financial Statistics* (lines 34 + 35 for M2); Philippines, *CB Statistical Bulletin*, various issues.

changed after 1965, when the demand for funds considerably increased, owing partly to brisk economic activity and partly to the aggressive rediscounting policy of the Central Bank, the government authorities did not amend the Usury Act, apparently to protect their overall economic policy of stimulating investment, particularly in import-substituting industries, through lower interest rates. The extent of the difference between the statutory and market rates is shown in table 12.

The low deposit and lending rates led to an increasing gap between supply of and demand for loanable funds. Since the rediscounting policy of the Central Bank was relatively lenient and the spread between rediscount and loan rates was very attractive, banks tended to rely more heavily on the Central Bank loans for relending to cover the gap. Banks, therefore, were becoming more active as conduits of Central Bank funds and less so as intermediaries between savers and borrowers.

The low interest rate and rediscount policy of the Central Bank also affected banks' loan portfolio. As may be gathered from table 13, bank loans tended to be concentrated at the shorter end. The reason for this is twofold: one is that banks could earn more by lending short at the prevailing rate and rolling over the loans several times, and the other is that only short-term papers were eligible for rediscounting.

The low interest-rate policy had retarded the development of government securities. In the 1950s, a very large share of total outstanding government securities went into the hands of the Central Bank (see table 14). Private wealth holders shied away from government securities because of their unattractive yields. The Central Bank therefore had to absorb them because of its standing policy of buying bonds offered for sale in order to make these bonds liquid. According to Valenzona,[7] "this policy has furnished a ready buyer for any government security offered for sale at par thus discouraging the trading of government securities." In the 1960s, however, private financial institutions tended to hold more government securities not because they had attractive yields, but rather because they were made eligible as reserves against deposit liabilities.

The underdevelopment of the equity market can also be traced to the low

Table 11. Total loans granted by commercial banks by interest rate, 1960–1973 (million pesos)

Period	Total	Interest rate percentage										
		4[a]	5	6	7	8	9	10	11	12	13	14
1960	2,605.8	32.2	35.9	320.4	864.1	679.3	352.3	213.3	3.0	105.6	—	—
1961	3,889.4	30.7	24.5	533.7	1,476.9	780.2	587.2	298.9	3.0	154.6	—	—
1962	4,773.1	71.2	32.9	657.0	1,045.9	877.2	1,174.4	585.5	61.0	228.6	—	—
1963	6,601.1	70.0	109.4	863.2	801.4	2,069.1	1,374.0	792.6	175.0	346.4	—	—
1964	7,123.8	37.9	27.9	482.4	979.7	1,894.1	1,099.9	1,029.5	339.9	530.7	—	1.8
1965	7,628.5	44.1	2.7	314.6	1,049.5	1,356.6	2,374.7	1,179.1	479.3	787.0	16.7	24.2
1966	8,050.8	26.0	5.1	456.6	469.0	505.4	1,915.6	1,996.1	1,195.3	1,171.3	129.0	181.4
1967	9,630.0	22.2	74.5	800.0	386.3	537.2	2,228.2	2,592.9	1,236.7	1,425.5	119.4	207.1
1968	12,428.4	24.9	122.2	309.4	1,168.6	545.8	1,514.7	3,695.5	2,834.3	1,682.9	285.7	244.4
1969	13,595.1	58.4	23.4	125.7	1,087.2	526.0	1,267.3	3,757.6	3,405.2	2,410.4	493.5	440.4
1970	16,827.0	64.4	47.7	128.5	260.4	303.8	1,668.4	2,239.5	5,228.6	4,520.4	1,300.4	1,064.9
1971	22,817.2	62.2	28.3	92.8	82.8	218.7	984.5	2,051.3	6,660.2	6,995.8	2,482.2	3,158.4
1972	26,948.1	235.3	52.2	256.4	223.3	338.9	875.4	2,290.5	6,885.1	7,798.0	2,368.5	5,654.5
1973	37,226.0	2,689.4	444.4	521.0	690.7	699.3	1,114.8	1,944.8	9,901.6	9,895.4	2,477.5	6,847.1

a. Inclusive of 0 to 4.5 per cent.
Source: CB Statistical Bulletin, various issues.

Table 12. Comparison of statutory and market lending rates

	Statutory rates (% per annum)	Effective rates (% per annum)
Commercial banks	9–14	12.18–16.78
Rural banks	12–14	15–18
Development banks	12	15
Investment banks	9–12	13–15
Government-financed institutions	9–12	14–16
Insurance firms	12–14	28–32
Consumer finance	12–14	45–60
Unregulated markets		60–400
Commercial paper	9.75–11.75	—
Government securities:		
Short-term bills	10–14	16–18
Medium-term notes	7–10	—
Long-term bonds	7	—

Source: *Report of the Inter-Agency Committee on the Study of Interest Rates* (1971).

Table 13. Total credits granted by commercial banks, classified by maturity (million pesos)

Period	Demand	Short-term[a]	Intermediate-term[b]	Long-term[c]	Total
1. Low interest rate period					
1960	667	2,050	128	4	2,849
	(23.4)[d]	(72.0)	(4.5)	(0.1)	(100.0)
1961	1,100	2,924	92	17	4,133
	(26.6)	(70.7)	(2.2)	(0.4)	(100.0)
1962	768	4,021	95	33	4,917
	(15.6)	(81.8)	(1.9)	(0.7)	(100.0)
1963	1,001	5,570	191	64	2,826
	(14.7)	(81.6)	(2.8)	(0.9)	(100.0)
1964	1,053	6,063	132	102	7,350
	(14.3)	(82.5)	(1.8)	(1.4)	(100.0)
1965	1,000	6,591	159	17	7,767
	(12.9)	(84.8)	(2.0)	(0.2)	(100.0)
1966	1,231	6,682	214	38	8,165
	(15.1)	(81.8)	(2.6)	(0.5)	(100.0)
1967	1,380	8,134	210	30	9,754
	(14.1)	(83.4)	(2.2)	(0.3)	(100.0)
1968	2,799	12,319	196	22	15,336
	(18.2)	(80.3)	(1.3)	(0.1)	(100.0)
1969	3,494	12,730	111	55	16,390
	(21.3)	(77.7)	(0.7)	(0.3)	(100.0)
1970	4,701	16,929	248	74	21,952
	(21.4)	(77.1)	(1.1)	(0.3)	(100.0)

Table 13 (*continued*)

Period	Demand	Short-term[a]	Intermediate-term[b]	Long-term[c]	Total
1971	5,692	22,446	530	153	28,821
	(19.7)	(77.9)	(1.8)	(0.5)	(100.0)
1972	6,867	24,720	757	345	32,689
	(21.0)	(75.6)	(2.3)	(1.1)	(100.0)
1973	9,089	35,457	827	1,312	46,685
	(19.5)	(75.9)	(1.8)	(2.8)	(100.0)

2. Transition period

1974	22,077	54,849	2,373	3,004	82,303
	(26.8)	(66.6)	(2.9)	(3.6)	(100.0)
1975	41,010	71,607	1,058	691	114,366
	(35,9)	(62.6)	(0.9)	(0.6)	(100.0)
1976	39,025	90,435	918	1,346	131,724
	(29.6)	(68.6)	(0.7)	(1.0)	(100.0)

Loans outstanding of commercial banks classified by maturity[e]

1977	7,490	26,767	4,234	1,683	40,173
	(18.6)	(66.6)	(10.6)	(4.2)	(100.0)
1978	9,164	35,227	5,549	4,139	54,078
	(16.9)	(65.1)	(10.3)	(7.6)	(100.0)
1979	10,637	37,601	9,389	10,637	68,264
	(15.6)	(55.1)	(13.7)	(15.6)	(100.0)
1980	10,458	49,844	7,747	9,149	77,198
	(13.5)	(64.6)	(10.0)	(11.8)	(100.0)

3. Floating interest rate period

1981	10,677	52,823	14,976	8,038	86,505
	(12.3)	(61.1)	(17.3)	(9.3)	(100.0)
1982	9,308	58,478	17,778	12,676	98,240
	(9.5)	(59.5)	(18.1)	(12.9)	(100.0)
1983	10,434	66,792	16,858	17,304	111,388
	(9.4)	(60.0)	(15.1)	(15.5)	(100.0)
1984[f]	10,624	72,484	21,035	16,880	121,023
	(8.8)	(59.9)	(17.4)	(13.9)	(100.0)

a. Loans with maturities of one year or less.
b. Loans with maturities of more than one year but less than five years.
c. Loans with maturities of more than five years.
d. Figures in parentheses are percentages.
e. Beginning in 1977, the series for total credits granted by commercial banks classified by maturity was discontinued.
f. As of September 1984.
Source: Department of Economic Research, Central Bank of the Philippines, *Central Bank Statistical Bulletin* (1983 and 1984 figures).

Table 14. Holders of outstanding government securities (million pesos)

Year	Total	Central Bank	Commercial banks	Savings and other banks[a]	Trust funds	Semi-govt. entities	Private sector	Foreign holders	Growth rate of total
1. Low interest rate period									
1957	1,108	838 (75.6)[b]	100 (9.0)	—	82 (7.4)	70 (6.3)	18 (1.6)	—	—
1958	1,300	1,024 (78.8)	88 (6.8)	—	101 (7.8)	66 (5.1)	21 (1.6)	—	17.3
1959	1,391	1,048 (75.3)	85 (6.1)	—	134 (9.6)	104 (7.5)	20 (1.4)	—	7.0
1960	1,484	1,027 (69.2)	106 (7.1)	—	175 (11.8)	147 (9.9)	29 (2.0)	—	6.7
1961	1,885	1,075 (57.0)	412 (21.9)	—	204 (10.8)	167 (8.9)	27 (1.4)	—	27.0
1962	1,885	1,083 (57.4)	349 (18.5)	—	242 (12.8)	182 (9.7)	29 (1.5)	—	0.0
1963	1,989	963 (48.4)	429 (21.6)	—	283 (14.2)	274 (13.8)	40 (2.0)	—	5.5
1964	2,076	1,058 (51.0)	378 (18.2)	—	320 (15.4)	277 (13.3)	43 (2.1)	—	4.4
1965	2,316	1,000 (43.2)	451 (19.5)	—	387 (16.7)	366 (15.8)	53 (2.3)	59 (2.5)	11.6
1966	2,675	1,238 (46.3)	644 (24.1)	—	366 (13.7)	261 (9.8)	111 (4.1)	55 (2.0)	15.5
1967	3,231	1,507 (46.6)	906 (28.0)	—	450 (13.9)	196 (6.1)	121 (3.7)	51 (1.6)	20.8

Year									
1968	3,594	1,610 (44.8)	1,034 (28.8)	—	494 (13.7)	228 (6.3)	181 (5.0)	47 (1.3)	11.2
1969	4,847	2,225 (45.9)	1,396 (28.8)	—	594 (12.3)	272 (5.6)	317 (6.5)	43 (0.9)	34.9
1970	5,233	2,385 (45.6)	1,334 (25.5)	—	685 (13.1)	246 (4.7)	542 (10.4)	41 (0.8)	8.0
1971	5,825	2,422 (41.6)	1,595 (27.4)	147 (2.5)	794 (13.6)	244 (4.2)	584 (10.0)	39 (0.7)	11.3
1972	7,647	3,022 (39.5)	1,646 (21.5)	125 (1.6)	865 (11.3)	329 (4.3)	1,630 (21.3)	30 (0.4)	31.3
1973	11,203	3,571 (31.9)	3,906 (34.9)	285 (2.5)	908 (8.1)	1,045 (9.3)	1,458 (13.0)	29 (0.3)	46.5
2. Transition period									
1974	14,989	4,190 (28.0)	4,142 (27.6)	258 (1.7)	1,043 (7.0)	2,399 (16.0)	2,934 (19.6)	23 (0.2)	33.8
1975	19,567	4,426 (22.6)	5,703 (29.1)	340 (1.7)	1,433 (7.3)	4,108 (21.0)	2,297 (11.7)	1,260 (6.4)	30.5
1976	22,093	4,764 (21.6)	6,329 (28.6)	444 (2.0)	1,682 (7.6)	5,668 (25.7)	1,949 (8.8)	1,257 (5.7)	12.9
1977	25,787	5,153 (20.0)	8,390 (32.5)	474 (1.8)	1,699 (6.6)	6,540 (25.4)	2,343 (9.1)	1,188 (4.6)	16.7
1978	29,001	4,889 (16.9)	11,026 (38.0)	704 (2.4)	2,119 (7.3)	5,889 (20.3)	3,195 (11.0)	1,179 (4.1)	12.5
1979	31,434	4,958 (15.8)	11,616 (37.0)	760 (2.4)	2,701 (8.6)	8,031 (25.5)	2,826 (9.0)	542 (1.7)	8.4
1980	34,262	6,879 (17.7)	11,965 (34.9)	100 (2.9)	4,327 (12.6)	6,990 (20.4)	3,635 (10.6)	262 (0.8)	9.0

Table 14 (continued)

Year	Total	Central Bank	Commercial banks	Savings and other banks	Trust funds	Semi-govt. entities	Private sector	Foreign holders	Growth rate of total
3. Floating interest rate period
1981	39,467	6,424 (16.3)	13,475 (34.1)	977 (2.5)	4,936 (12.5)	9,561 (24.2)	3,832 (9.7)	262 (0.7)	15.2
1982	46,975[c]	9,168 (19.5)	15,213 (32.4)	1,019 (2.2)	5,280 (11.2)	10,917 (23.2)	5,116 (10.9)	262 (0.6)	19.0
1983	50,818[c]	7,841 (15.4)	18,275 (36.0)	—	5,748 (11.3)	13,554 (26.7)	4,810 (9.5)	590 (1.2)	8.2
1984	n.a.	n.a.	n.a.	n.a.	n.a.	n.a.	n.a.	n.a.	—

a. Starting in 1970, savings and other banks became a component of the Thrift Bank.
b. Figures in parentheses are percentages of total.
c. Exclusive of Treasury Bills.
Source: Central Bank Statistical Bulletin, various issues.

interest-rate policy during the period 1956–1973, and to the highly fragmented financial system. With a net rate of return on equity of between 17 and 18 per cent per annum, it was less costly for firms to borrow from banks who were charging the rates prescribed by the Usury Act rather than to raise equity by selling stocks. Besides, by not selling shares, owners, especially those of closely held family enterprises, would not lose control or management of their firms to new stockholders. Data on debt–equity ratios seem to support this view (see table 15). All industries show quite high debt–equity ratios except that of the mining industry. Most of the mining firms had shares traded in the stock market because of their relatively large capital requirement.

Another reason why firms did not go to the equity market to raise funds was the relatively underdeveloped investment banking sector. Investment banks were small and concentrated mainly in retailing rather than wholesaling of securities. This raised unnecessarily the underwriting cost. In her study, Valenzona[8] noted that the volume of securities placed and floated by the investment banking sector constituted a very small percentage of the total volume of securities held (see table 16). The commission rates on flotation and underwriting of securities ran between 2 and 5 per cent of the total value of the shares marketed.

Big commercial banks and insurance firms were in a better position to do wholesaling of securities at lower cost, but the Investment Company Act prohibited them from going into investment banking, presumably to protect them from the riskiness of the securities market.

The rigid and low interest-rate policy could not accommodate the needs of a growing economy. Individuals and institutions with temporary surplus units were forced to hold either cash or deposits with low returns. In the mid-1960s, the emergence of new financial institutions, called non-banks, which were not directly under the control of the Central Bank, paved the way for the development of the money market. This offered a variety of short-term financial instruments with unregulated rates that sometimes went beyond 30 per cent on an annual basis, and it inevitably drew resources away from traditional deposits. It could have caused sudden disintermediation and, therefore, the collapse of the banking system, had the latter not responded to the challenge posed by non-bank financial institutions by also offering deposit substitutes whose rates were unregulated.

The growth of the money market was phenomenal. The volume of money-market transactions by investment houses rose from 32 million pesos in 1966 to 1,756 million pesos in 1973, and by commercial banks from 954 million pesos in 1966 to 6,202 million pesos in 1973.[9] The total short-term deposit substitutes outstanding in the commercial banking system amounted to 60 per cent of its total medium- and long-term time and savings deposits in 1973.

Although the money market undermined the growth of traditional deposits during the low interest-rate period, nevertheless it "played a vital role in the last few years in mobilizing a larger share of incomes in the form of financial

Table 15. Debt equity ratio

Earlier industry classification	1968	1969	1970	1971	1972	1973	1974	1975	1976	1977	1978	Present industry classification	1979	1980	1981	1982	1983
Agriculture	2.88	2.54	1.91	2.45	2.69	2.48	2.19	1.61	2.04	1.62	2.02	Agriculture, fishery, and forestry	1.75	2.07	2.40	2.15	4.18
Manufacturing	1.38	1.23	1.56	1.67	1.80	1.69	2.44	1.77	1.82	1.90	2.16	Manufacturing	2.17	2.51	2.17	2.13	2.43
Mining	1.83	0.54	0.59	0.48	0.53	0.66	0.82	1.02	1.19	1.42	1.59	Mining and quarrying	0.88	1.14	2.09	3.47	4.26
Services	0.94	0.81	1.20	1.53	1.53	1.30	1.69	1.94	1.30	1.98	2.04	Construction	3.00	2.98	3.49	4.15	2.40
Utilities	2.23	1.64	2.54	2.64	2.86	2.47	2.25	2.09	2.31	2.27	1.97	Electricity, gas, and water	1.21	0.97	1.32	1.42	1.57
Commercial	1.78	1.66	1.96	1.97	1.96	2.05	2.42	2.50	3.54	2.41	2.38	Transportation, storage, and communication	2.47	2.79	3.16	3.69	6.50
Unclassified	2.24	1.24	1.90	1.90	0.40	0.47						Wholesale and retail trade	3.04	3.50	3.92	3.97	3.79
												Community, social, and personal services	2.26	1.56	1.67	1.78	1.12
												Financing, insurance, real estate, and business services	8.16	0.30	10.46	9.65	10.35
All	1.53	1.23	1.56	1.64	1.72	1.61	2.05	1.77	1.94	1.99	2.09	All	3.64	3.96	4.65	4.66	5.03

Source: *Top 1000 Corporations, Business Day*, various issues.

Table 16. Primary sales of securities of the investment banking sector 1963–1967 (millions of pesos)

Year	Primary sales	Private placements	Private placements Primary sales (%)
1963	5.0	No data	—
1964	3.0	No data	—
1965	16.0	6.0	37.5
1966	33.2	18.0	54.2
1967	26.3	11.3	43.0

Source: Valenzona (1969), table II.4.

assets."[10] The growing sophistication of the financial system required reforms; these were introduced in 1972 and included the following.[11]

1. No more commercial banks were to be added, and the maximization of efficiency in the existing units was made the subject of current concern: i.e. branch banking was given preference over unit banking.
2. The classification of banking institutions was simplified and reduced from five different types of institutions to three main categories. The application of monetary and fiscal policies was to be made uniform for institutions rendering similar services or enjoying similar benefits.
3. The Central Bank was given authority and responsibility not only over the monetary system, but over the entire financial and credit system as well.
4. The responsibility of the Central Bank was redefined primarily as the maintenance of monetary stability, while the responsibility for promoting growth was to be shared with the planning agencies of the government.
5. The Central Bank was given more flexibility in exercising powers consistent with the maintenance of monetary stability.

It is to be noted that the 1972 financial reforms enforced specialization among various types of financial entities. Specifically, "investment banking activities" were separated from "regular banking activities." The former were reserved solely to investment houses, which were established under the Investment Houses Law of 1973.

In 1973, the Monetary Board was given the authority to prescribe the maximum lending rates, thus virtually repealing the Usury Act of 1916.

Some of the aspects of the 1972 financial reforms are discussed in the following subsection.

Transition Period, 1974–1980

The transition period marks the beginning of financial liberalization. Financial policies were revised in order to mobilize long-term funds for investment and

Table 17. Interest-rate reforms, 1974–1979

Instruments	Before 1974	1974	1976	1977	1979
Savings deposits	6	6	7	7	9
Time deposits					
Short-term	6.5–8.0	8.0–11.0	8.5–12.0	8.5–12.0	10.5–14.0
Long-term	6.5–8.0	No ceiling	No ceiling	No ceiling	No ceiling
Deposit substitutes					
Short-term		Unregulated	17	15 (effective)	17 (effective)
Long-term		Unregulated	No ceiling	No ceiling	No ceiling
Loans					
Short-term[a]					
Secured	12	12	12	12 (effective)[c]	14 (effective)[c]
Unsecured	14	14	14	14 (effective)[c]	16 (effective)[c]
Long-term[b]					
Secured	12	12	19	19 (effective)[c]	21 (effective)[c]
Unsecured	14	14	19	19 (effective)[c]	21 (effective)[c]

a. Short-term: 730 days or less.
b. Long-term: more than 730 days.
c. Excludes other charges of up to 2 or 3 per cent. Before 1977, the interest rate on loans equals the basic rate plus other charges. From 1977 to 1979, the effective interest rate on loans refers to the basic rate only.

Source: Various Central Bank circulars.

develop the capital market. During this period, interest rates were still administratively fixed but were constantly adjusted by the Monetary Board to reflect market conditions.

The Monetary Board wasted no time in exercising its newly acquired power by initiating five interest-rate reforms within a span of seven years (see table 17). In 1974, ceilings on short-term deposits were increased from 6.5–8.0 per cent to 8.0–9.5 per cent, while the rate on long-term time deposits was deregulated. This was the first time that the deposit instruments of longer maturities were not subjected to ceilings. The rates of other instruments remained the same, however. This clearly manifests the serious effort of the monetary authorities to mobilize more long-term funds.

Another interest-rate reform was initiated in 1976. This time, the reform was intended to cope with the rapidly growing deposit substitutes and other high-yielding, short-term money-market instruments. The interest ceiling on savings deposits was increased to 7 per cent, and that on time deposits to 8.5–12.0 per cent, making the interest differential between short- and long-term deposits wider. At the same time, an interest ceiling of 17 per cent was imposed on short-term deposit substitutes. In addition, the minimum placement on deposit substitutes was increased to 200,000 pesos for maturities of 730 days or less and 100,000 pesos for maturities of more than 730 days. Furthermore, a reserve requirement of 20 per cent on the deposit substitutes of commercial banks and non-bank financial institutions was introduced. It is noteworthy that reforms were intended to reverse the flow of funds from short-term to long-term financial instruments. The 1976 interest-rate reform introduced for the first time the distinction between short-term and long-term loans. The former was subjected to the old ceilings, while the latter carried a ceiling of 19 per cent per annum.

The interest-rate reform of 1977 was intended to reduce the effective cost of funds to borrowers to stimulate investment. Thus, the interest-rate ceilings on loans were made the effective interest-rate ceilings; this saved borrowers from paying other charges imposed by banks on loans. Aside from this, the effective ceiling on short-term deposit substitutes was reduced to 15 per cent. Also, a 35 per cent transactions tax was imposed on all primary borrowings in the money market. This further reduced the interest-rate differential between money-market instruments and traditional deposits.

In 1979, the rising interest rate in the international capital market and the stronger inflationary pressures felt by the domestic economy owing to the second oil shock prodded the Monetary Board to increase the interest-rate ceilings on various instruments by 2 percentage points across the board. The increase in interest ceilings was deemed sufficient to defend the interest incomes of savers and lenders from being eroded by high inflation and the unnecessary outflow of capital due to high interest rates abroad.

To set the tone for a fully floating interest rate, the Monetary Board deregulated the rate on loans with maturities of four years or more, beginning in 1980.

Although interest rates were adjusted upwards, the interest-rate reforms during the transition period still retained some basic ingredients from the earlier Central Bank's interest-rate policy of making the cost of credit affordable to potential borrowers. Thus, interest rates on deposits could not be adjusted to a level reflecting market conditions, since interest ceilings were imposed on loans.

The effort to liberalize was also extended to other areas of the financial system. A significant part of the 1972 financial reforms was concerned with the raising of the minimum paid-in capital of commercial banks from 20 to 100 million pesos, in order to rationalize the banking system and to increase bank resources available for long-term loans. To satisfy the new requirement, mergers and consolidation among banks were encouraged. Apart from this, equity participation of foreign entities in domestic banks of up to 30–40 per cent was allowed.

Several banks did merge to satisfy the new minimum capital requirement and also to position themselves in an increasingly competitive environment. The

Table 18. Foreign investors in Philippine financial institutions

	Foreign investor	Equity participation (%)
Commercial banks		
BPI	Morgan Guaranty Trust Co.	20.4
Comtrust	Chase Manhattan Bank	20.0
FEBTC	Chemical International Finance of New York	12.7
Rizal Commercial Banking Corp.	Continental International Finance Corp.	30.0
Security Bank and Trust Co.	Bank of Nova Scotia	30.0
Traders Royal Bank	Royal Bank of Canada	30.0
Feati	First National City Bank of New York	30.0
General	Grindlays Bank Ltd.	31.2
City Trust Bank	Citibank	43.0
Investment houses		
Ayala Investment and Development Corp.	Wells Fargo Bank International	N.a.
State Investment	Hongkong and Shanghai Banking Corp.	N.a.
BANCOM Development Corp. (merged with Union Savings Bank)	American Express Co.	N.a.

mergers resulted in the reduction in the number of domestic commercial banks from 34 to 27. Equally significant was the favourable response of foreign entities to the new policy environment. Eight commercial banks and nine investment houses received substantial foreign capital infusion (see table 18). Aside from shoring up the capital of domestic financial institutions, the participation of foreign entities also improved the managerial capabilities of local bank personnel and exposed domestic banks to the opportunities available in the international capital market.

The financial reforms also included provisions allowing commercial banks, for the first time, to invest in the equities of non-allied financial undertakings, such as warehousing, leasing, storage, insurance, home building, etc. The idea was to strengthen these relatively important economic activities by ensuring a flow of long-term capital to them.

The establishment of expanded foreign currency deposit units (FCDUs) and offshore banking units (OBUs) in 1976 represents another major effort at financial liberalization.[12] The development of the foreign currency deposit system was "aimed primarily at widening the access of the country to foreign currency funds, gaining familiarity with foreign investors, and exposing domestic corporations to the foreign market, thereby enabling the country to compete in the international capital markets."[13]

Although directly supervised by the Central Bank, FCDUs and OBUs, however, operate in a fairly liberal environment. They are exempted from the interest-rate ceilings and the 5 per cent gross receipts tax. They are required to pay taxes equivalent only to 5 per cent for their net offshore income and 10 per cent of their gross onshore income. They can solicit foreign currency deposits from residents (resident commercial banks only for OBUs) and non-residents, and lend to bank and non-bank residents and/or non-residents. The specific regulations covering the operations of 343 banks,[14] FCDUs, and OBUs are summarized in table 19.

This new policy received a favourable response from potential participants. As of December 1980 29 domestic banks, including four branches of foreign banks and two government-owned banks, were given authority to operate foreign currency deposit units, while 21 foreign banks were authorized to operate offshore banking units.

As mentioned earlier, the interest-rate reforms initiated during the transition period were mainly aimed at reducing and stabilizing the "money-market rates and providing a base for the development of long-term markets."[15] The reforms gave some impressive results. Specifically, the share of banks' time deposits in total deposit liabilities was steadily increasing during the transition period at the expense of deposit substitutes and demand deposits (see table 20). Furthermore, the share of longer-term time deposits increased as a proportion of total time deposits. This could be the result of increasing the interest-rate differential between shorter-term and longer-term time deposits. The nominal growth rate of banks' time deposits during the transition period averaged 43 per cent annually, compared with only 25 per cent annually during the low interest-rate period. Interestingly, the growth of banks' deposit substitutes decelerated from 85 per cent in 1974 to only 3.5 per cent in 1980.

It is worth while to refer again to table 9, which provides a convenient summary of the impact of interest-rate reforms on the flow of loanable funds during the transition period. As may be gathered from the table, the reforms made possible the steady rise of M3/GNP ratio from 1974 to 1978. During this period, the real interest rate on time deposits and deposit substitutes was positive. However, the ratio started to decline in the succeeding years, 1979 and 1980, when real interest rates on both traditional deposits and deposit substitutes were severely negative as a result of the high inflation caused by the second oil shock. Again, we observe disintermediation occurring at times when real in-

Table 19. The three-tiered foreign currency deposit system

	343 Banks	Foreign Currency Deposit Units (FCDUs)	Offshore Banking Units (OBUs)
Regulations governing operations	RA 6426 CBC 343	PD 1035 CBC 547 CBC 623 Revenue Regulations 10/76 CBC 685	PD 1034 CBC 546 Revenue regulations 10/76 CBC 685
Participating banks	Savings banks; Commercial banks (Philippine commercial banks plus the 4 branches of foreign banks)[a]	Commercial banks	Branches of foreign banks
Qualified depositors	Residents Non-residents	Residents Non-residents	Resident commercial banks only
Amount of deposit	Any amount	Any amount	For time/call deposits: $50,000 and above For demand deposits: any amount
Types of deposits	Savings Time Demand Trust	Savings Time Demand Trust	Time/call Demand (There are no savings accounts or regulations as to trust deposits)

Currencies used	Pesos Reserve currencies	Pesos Reserve currencies Any currency freely convertible into the reserve currencies	Any foreign currency. (For transactions with residents only reserve currencies or any foreign currency freely convertible into reserve currencies. The peso is used for administrative transactions only)
Liquidity ratios	For every deposit, there exists a 100% foreign currency cover requirement, 15% of which is a mandatory deposit with the Central Bank. This takes the place of the reserve requirement[b]	For every deposit, there exists a 100% foreign currency cover requirement, but a deposit with the Central Bank is *not* required. The option to deposit 100% abroad is open	There is no provision on the cover. All that is required is an unconditional guarantee by the head office which suggests, in effect, a 100% cover
Currency trading	No foreign exchange trading is allowed; the 100% cover is required to be in the same currencies as they were when deposited.	Foreign exchange trading is allowed, but at the end of the day the bank is to retain at least 70% of the cover in the original currency; 30% may be in other currencies	Foreign exchange trading is allowed with no restrictions on what currency the bank is to hold
Minimum capitalization requirement	For savings banks – 5 million pesos For commercial banks – 20 million pesos	150 million pesos	No capitalization requirement. (The law simply requires that they hold at least US$1 million in deposits with the Central Bank or investments in Philippine securities or other CB-approved assets)

Table 19 (continued)

	343 Banks	Foreign Currency Deposit Units (FCDUs)	Offshore Banking Units (OBUs)
Securities	Short-term and readily marketable (i.e. 1 year)	Short-term Medium-term Long-term	Short-term Medium-term Long-term
Qualified borrowers	Qualified residents (BOI-registered and CB-certified export-oriented firm)	Residents Non-residents (subject to Central Bank approval)	Residents (subject to Central Bank approval) Non-residents (no Central Bank approval needed)
Swap operations (foreign currency in pesos)	100% (the bank can swap all foreign currencies to pesos and then relend them to domestic borrowers[c] Swap with CB only	100% (the bank can swap all foreign currencies to pesos and then relend them to domestic borrowers[c] Swap with CB, FCDU, OBU (with OBU, can swap in foreign currencies only)	0% (OBUs do not have access to pesos, thus there are no swap arrangements in pesos) Swap with CB, FCDU, OBU (only in foreign currencies)
Uniform Currency Law (RA 529)/Usury Law (RA 2655)	Applicable	Not applicable	Not applicable

FINANCIAL LIBERALIZATION: THE PHILIPPINE CASE   237

Taxation:			
– 5% gross receipts tax (Tax Code, sec. 249)	Applicable	Not applicable	Not applicable
– Documentary and science tax	Applicable	Not applicable	Not applicable
– Tax on net income	For income above 100,000 pesos the tax rate is 35%; for income below 100,000 pesos the tax rate is 25%	For net offshore income, the tax rate is 5%; for gross onshore income, the tax rate is 10%	For net offshore income, the tax rate is 5%; for gross onshore income, the tax rate is 10%

a. These are: Bank of America, First National City Bank (Citibank), the Hongkong and Shanghai Banking Corporation, and the Chartered Bank. These banks were the only four branches of foreign banks which were already in operation before the ban on the establishment of foreign bank branches in the Philippines was imposed.
b. The Central Bank interest rate is equal to the lower of SIBOR or LIBOR, or it may be between these two rates. This restriction on 15 per cent of the cover suggests that the return on investment may be much less, since alternative uses (deposits abroad, domestic loans, etc.) may earn a higher return. Only 85 per cent may be invested in these alternative uses.
c. Swap operations allow the bank to earn a higher return on investment, since pesos are more expensive than foreign currency. For foreign currency–peso swaps, FCDUs may transact business either with the Central Bank or other FCDUs.
Source: Africa (1980), pp. 29–35.

Table 20. Structure of deposits and bills payable of commercial banks (million pesos)

End of year	Deposits				Deposit substitutes	Bills payable		Grand total
	Demand	Savings	Time	Subtotal		Other BP	Subtotal	

1. Low interest rate period (1957–1973)

End of year	Demand	Savings	Time	Subtotal	Deposit substitutes	Other BP	Subtotal	Grand total
1957	709.4	500.5	118.8	1,328.7	—	121.7	121.7	1,450.4
	(48.9)[a]	(34.5)	(8.2)	(91.6)		(8.4)	(8.4)	(100.0)
1958	740.7	567.8	144.0	1,452.5	—	125.5	125.5	1,578.0
	(46.9)	(36.0)	(9.1)	(92.0)		(8.0)	(8.0)	(100.0)
1959	775.5	634.2	183.4	1,593.1	—	144.9	144.9	1,738.0
	(44.6)	(36.5)	(10.6)	(91.7)		(8.3)	(8.3)	(100.0)
1960	774.9	714.7	242.2	1,732.0	—	81.2	81.2	1,813.2
	(42.7)	(39.4)	(13.4)	(95.5)		(4.5)	(4.5)	(100.0)
1961	972.8	895.2	481.3	2,349.3	—	275.6	275.6	2,624.9
	(37.1)	(34.1)	(18.3)	(89.5)		(10.5)	(10.5)	(100.0)
1962	1,152.2	988.6	863.1	3,003.9	—	117.5	117.5	3,121.4
	(36.9)	(31.7)	(27.6)	(96.2)		(3.8)	(3.8)	(100.0)
1963	1,357.7	1,164.5	1,125.2	3,647.4	—	322.7	322.7	3,970.1
	(34.2)	(29.3)	(28.3)	(91.9)		(8.1)	(8.1)	(100.0)
1964	1,259.8	1,313.9	1,036.6	3,610.3	—	728.9	728.9	4,339.2
	(29.0)	(30.3)	(23.9)	(83.2)		(16.8)	(16.8)	(100.0)
1965	1,545.7	1,402.5	1,053.9	4,002.1	—	814.1	814.1	4,816.2
	(32.1)	(29.1)	(21.9)	(83.1)		(16.9)	(16.9)	(100.0)
1966	1,521.4	1,925.4	1,215.2	4,662.0	—	953.9	953.9	5,615.9
	(27.1)	(34.3)	(21.6)	(83.0)		(17.0)	(17.0)	(100.0)
1967	1,756.5	2,455.5	1,416.8	5,628.8	—	1,605.9	1,605.9	7,234.7
	(24.3)	(33.9)	(19.6)	(77.8)		(22.2)	(22.2)	(100.0)

FINANCIAL LIBERALIZATION: THE PHILIPPINE CASE

Year							
1968	1,861.6	2,762.0	1,310.2	5,933.8	—	1,778.3	7,712.1
	(24.1)	(35.8)	(17.0)	(76.9)		(23.1)	(100.0)
1969	2,372.2	3,127.9	1,258.7	6,758.8	—	1,991.4	8,750.2
	(27.1)	(35.7)	(14.4)	(77.2)		(22.8)	(100.0)
1970	2,458.9	3,757.2	1,469.6	7,685.7	—	1,948.7	9,634.4
	(25.5)	(39.0)	(15.2)	(79.8)		(20.2)	(100.0)
1971	2,915.9	4,409.5	1,889.5	9,214.9	—	2,233.8	11,448.7
	(25.5)	(38.5)	(16.5)	(80.5)		(19.5)	(100.0)
1972	3,735.4	4,669.7	2,558.9	10,964.0	—	3,082.9	14,046.9
	(26.6)	(33.2)	(18.2)	(78.1)		(21.9)	(100.0)
1973	5,276.2	6,865.4	3,044.2	15,185.8	4,041.4	1,789.1	21,016.3
	(25.1)	(32.7)	(14.5)	(72.2)	(19.2)	(8.5)	(100.0)
Average growth rate (%)	14.1	18.22	25.2			36.6	

2. Transition period (1974–1980)

Year							
1974	6,062.0	8,281.1	4,032.2	18,375.3	7,469.8	4,790.2	30,635.3
	(19.8)	(27.0)	(13.2)	(60.0)	(24.4)	(15.6)	(100.0)
1975	6,664.2	8,951.2	5,130.7	20,746.1	9,631.7	8,577.8	38,955.6
	(17.1)	(23.0)	(13.2)	(53.3)	(24.7)	(22.0)	(100.0)
1976	7,482.0	11,021.6	7,557.6	26,061.2	10,872.7	7,017.6	43,951.5
	(17.0)	(25.1)	(17.2)	(59.3)	(24.7)	(16.0)	(100.0)
1977	9,045.0	13,262.5	11,469.5	33,777.0	11,399.5	4,955.6	50,132.1
	(18.0)	(26.4)	(22.9)	(67.4)	(22.7)	(9.9)	(100.0)
1978	9,602.3	17,313.7	16,736.5	43,652.5	11,493.9	10,270.1	65,416.5
	(14.7)	(26.5)	(25.6)	(66.7)	(17.6)	(15.7)	(100.0)
1979	11,396.9	20,883.7	23,716.6	55,997.2	11,950.9	16,119.7	84,067.8
	(13.6)	(24.8)	(28.2)	(66.6)	(19.2)	(19.2)	(100.0)

Table 20 (*continued*)

End of year	Deposits				Bills payable			Grand total
	Demand	Savings	Time	Subtotal	Deposit substitutes	Other BP	Subtotal	
1980	12,792.0 (12.1)	23,947.1 (21.8)	36,791.1 (34.9)	72,630.2 (68.8)	12,371.4 (11.7)	20,510.8 (19.4)	3,288.2 (31.2)	105,512.4 (100.0)
Average growth rate (%)	13.6	19.09	43.06		19.96	55.8		
3. 1981–1984								
1981	12,197.9 (9.9)	24,837.2 (20.0)	41,493.4 (33.8)	78,528.5 (64.0)	15,922.6 (13.0)	28,182.3 (23.0)	44,104.9 (36.0)	122,633.4 (100.0)
1982	10,756.3 (9.1)	30,454.0 (25.9)	27,962.1 (23.8)	69,172.4 (58.8)	16,565.6 (14.1)	31,875.5 (27.1)	48,441.1 (41.2)	117,613.5 (100.0)
1983	19,139.5 (7.7)	40,939.7 (23.1)	56,147.8 (31.6)	116,227.0 (65.5)	17,106.3 (9.6)	44,033.5 (24.8)	61,139.8 (34.5)	177,366.8 (100.0)
1984	15,272.0 (7.7)	47,211.6 (23.8)	72,068.4 (36.4)	134,552.0 (68.0)	10,372.8 (5.2)	53,074.3 (26.8)	63,447.1 (32.0)	197,999.1 (100.0)
Average growth rate (%)	10.3	20.0	27.4		−0.8	27.3		

a. Figures in parentheses are percentages.
Source: *CB Statistical Bulletin*, various issues.

terest rates on financial assets became negative. Apparently, the upward adjustment in the interest rates in 1979 was not high enough to encourage wealth holders to hang on to their deposits.

In general, the interest-rate reforms improved the flow of loanable funds during the transition period: the M3/GNP ratio at this time averaged 26.6 per cent, compared with 23.2 per cent during the low interest-rate period. However, the Philippines' performance cannot compare with that of other countries. For example, the Republic of Korea's M2/GNP ratio remained at 33.7 per cent in 1980, whereas the Philippines' was only 25.6 per cent.

In the absence of a smoothly functioning securities market, the pressure to provide long-term funds for investment falls heavily on banks. Table 13 shows that the share of banks' intermediate- and long-term loans outstanding increased after the 1976 interest-rate reforms. Thus, the lengthening of the term structure of their deposit liabilities and the upward revision of interest-rate ceilings on long-term loans provided incentives to banks to increase the proportion of their medium- and long-term loans.

Government securities continued to dominate the securities market. Outstanding government securities increased from 15 billion pesos in 1974 to 34 billion pesos in 1980. Except for Treasury Bills, government securities had longer maturities, but they carried interest rates significantly below the market rate. During the transition period, commercial banks became the primary holders of government securities, simply because most of them, specifically CBCIs, were eligible as reserves against deposit liabilities and were allowed to be used as substitutes for the agricultural and agrarian reform credit requirements of commercial banks (see table 14). Others, like government institutions, held government securities mainly because of some regulations forcing them to invest in government securities. The private sector held only a small proportion of the total outstanding government securities. Thus, the government's pricing policy failed to create a broader market for its securities.

The performance of the private securities market during the transition period was no different from that during the low interest-rate period. It was generally believed that the administratively fixed lending rates during this period were still significantly below the market rates, so most private entities tended to rely more on borrowings. As may be gathered from table 15, the average debt–equity ratio for all industries continued to rise from 2.05 in 1974 to 3.96 in 1980. Also during this period, government financial institutions made an intensive effort to promote their lending operations: loans were made available to large borrowers at concessionary rates. Besides low interest rates, the prevailing tax system gave an undue advantage to incurring more debts rather than raising equity. This is because returns to equity are taxed, whereas interest expenses on loans cause the net taxable income to go down.

Investment houses, which were expected to strengthen the securities market after their establishment in 1973, showed a very disappointing performance. As

Table 21. Percentage distribution of assets and liabilities of investment houses (1975–1980)[a]

Item	1975	1976	1977	1978	1979	1980
Assets						
Cash items	3	5	8	9	11	10
Loans (net)	28	36	79	76	75	71
Investments	61	50	5	7	5	6
Other assets	9	9	8	8	9	13
Total assets	100	100	100	100	100	100
Liabilities and equity						
Borrowings	85	79	79	81	82	80
Short-term	70	65	64	68	66	N.a.
Long-term	15	14	15	13	16	N.a.
Other liabilities	5	10	8	5	5	8
Total liabilities	90	89	87	86	87	88
Total equity	10	11	13	14	13	12
Total liabilities and equation	100	100	100	100	100	100
Total resources						
Pesos (millions)	4,774	4,825	4,339	4,763	6,553	8,607
% change	N.a.	1	(10)	10	38	31

a. This table includes all investment houses, including PDCP. Details may not add up to total because of rounding.
Source: Licuanan (1986), table 11, p. 66.

table 21 shows, they tended to rely more heavily on short-term money market funds. On the asset side, investment houses reduced their exposure in investments in securities from 61 per cent of their total assets in 1975 to 6 per cent in 1980, and increased their loans, which were mostly of short-term maturity, from 28 per cent of total assets in 1975 to 71 per cent in 1980. The reliance of investment houses on asset-based rather than on fee-based activities, such as the underwriting of securities, made them more like ordinary banks than investment houses. It should be noted that providing short-term financing for corporate clients was not supposed to be their main activity. Licuanan[16] attributed this development to the depressed private securities market that forced investment houses to look for other opportunities to increase their earnings. But it may well be that their very limited capital base made them rely more on short-term bor-

Table 22. Philippine offshore banking and foreign currency deposit systems: Classification of assets by maturities (in US$ millions)

End of period	On demand	1–7 days	7 days to 1 year	1 to 5 years	Over 5 years	Total	Growth rate (%)
OB system							
1977	—	134	616	7	—	757	—
		(17.7)[a]	(81.4)	(0.9)		(100.0)	
1978	—	285	1,596	106	—	1,987	162.5
		(14.3)	(80.3)	(5.3)		(100.0)	
1979	—	326	2,449	169	—	2,944	48.2
		(11.1)	(83.2)	(5.7)		(100.0)	
1980	104	471	3,302	181	—	4,058	37.8
	(2.6)	(11.6)	(81.4)	(4.5)		(100.0)	
1981	151	487	3,716	110	163	4,627	14.0
	(3.3)	(10.5)	(80.3)	(2.4)	(3.5)	(100.0)	
1982	59	578	4,065	162	124	4,998	7.8
	(1.2)	(11.6)	(81.5)	(3.2)	(2.5)	(100.0)	
1983	70	169	3,792	175	202	4,408[b]	−11.6
	(1.6)	(3.8)	(86.0)	(4.0)	(4.6)	(100.0)	
1984	154	175	3,543	184	165	4,221[b]	−4.2
	(3.6)	(4.1)	(83.9)	(4.4)	(3.9)	(100.0)	
FCD system							
1980	1,127	295	2,858	929	—	5,209	—
	(21.6)	(5.7)	(54.9)	(17.8)		(100.0)	
1981	832	669	2,960	360	374	5,195	−0.3
	(16.0)	(12.9)	(57.0)	(6.9)	(7.2)	(100.0)	
1982	1,912	474	3,547	86	741	6,760	30.1
	(28.3)	(7.0)	(52.5)	(1.3)	(11.0)	(100.0)	
1983[c]	1,462	284	2,014	397	744	4,901	−27.5
	(29.8)	(5.8)	(41.1)	(8.1)	(15.2)	(100.0)	
1984[c]	1,295	198	2,776	52	92	4,413	−10.0
	(29.3)	(4.5)	(62.9)	(1.2)	(2.1)	(100.0)	

a. Figures in parentheses are percentages of total.
b. Includes allowance for probable losses.
c. Excludes Citibank's Due From/To Head Office, Branches, and Agencies Abroad Account, which amounts to $2,041 million, in both assets and liabilities.
Source: Foreign Exchange Department, Central Bank.

rowed funds, so that they also concentrated on short-term lending. Thus, the regulation stating that only investment houses could carry out investment bank-

Table 23. Onshore loan portfolios of FCDUs and OBUs (in US$ millions)

End of period	FCDUs			OBUs		
	Total	Loans to non-banks	Interbank transaction	Total	Loans to non-banks	Interbank transaction
1977	N.a.	N.a.	N.a.	375 (100.0)[a]	75 (20.0)	300 (80.0)
1978	N.a.	N.a.	N.a.	1,196 (100.0)	372 (31.1)	824 (68.9)
1979	2,713[b] (100.0)	2,054 (75.7)	659 (24.3)	1,972 (100.0)	855 (43.3)	1,117 (56.6)
1980	3,795 (100.0)	2,856 (75.2)	939 (24.7)	2,924 (100.0)	1,231 (42.1)	1,693 (57.9)
1981	3,817 (100.0)	2,660 (69.7)	1,157 (30.3)	3,444 (100.0)	1,567 (45.5)	1,877 (54.5)
1982	4,171 (100.0)	2,800 (67.1)	1,371 (32.9)	3,576 (100.0)	1,528 (42.7)	2,048 (57.3)
1983	3,403[b] (100.0)	2,775 (81.5)	628 (18.4)	3,325 (100.0)	1,495 (45.0)	1,830 (55.0)
1984	2,781[b] (100.0)	2,092 (75.2)	689 (24.8)	3,333 (100.0)	1,519 (45.6)	1,814 (54.4)

a. Figures in parentheses are percentages of total.
b. Excludes Citibank's Due From/To Head Office, Branches and Agencies Abroad Account, which amounts to $2,041 million, in both assets and liabilities.

Source: Foreign Exchange Department, Central Bank.

ing activities may also have undermined the development of the private securities market, since investment houses had only very limited resources compared to commercial banks.

The favourable policy environment provided by government to the offshore banking system produced impressive results.[17] In particular, resources of OBUs grew remarkably well from US$757 million in 1977 to US$4,058 million in 1980 (see table 22). The benefits accruing from the brisk operations of OBUs in the country are clear: they provided local borrowers with easier access to foreign currency funds. Table 23 shows that most OBUs' resources were lent onshore, either directly to non-bank customers or to domestic banks which in turn lent the funds to domestic non-bank customers. Thus, the inflow of foreign capital through OBUs supplemented the meagre supply of domestic capital. Moreover, loans of OBUs tended to have longer maturities. Whereas roughly 45 per cent of money transactions were on demand, only 10 to 18 per cent of OBUs loans outstanding had maturities of less than seven days. This means that relatively more funds were available for medium- and long-term lending. In table 22, we

Table 24. Central Bank deposits in OBUs (in US$ millions)

End of period	Total OBU liabilities (1)	Central Bank deposits w/OBUs (2)	2–1
1977	757	298	39.4
1978	1,987	440	22.1
1979	2,944	575	19.5
1980	4,058	503	12.4
1981	4,627	575	12.4
1982	4,988	751	15.0
1983	4,408	902	20.5
1984	4,221	949	22.5

Source: Foreign Exchange Department, Central Bank.

see that loans with maturities of over one year increased in absolute terms during the period 1977–1980. Thus, OBUs' contributions to the medium- and long-term capital funds available to domestic borrowers were not insignificant.

The presence of OBUs in the country also provided the Central Bank with an alternative outlet for its idle funds. In fact, the Central Bank's deposits with OBUs had been increased in absolute terms during the period considered in this study (see table 24). However, the Bank's deposits effectively increased OBUs' resources, which could have been used for onshore lending, and thus, its deposits with OBUs ultimately increased the capital available to domestic borrowers.

There are, however, several disadvantages to OBUs. One is that during the period of analysis they must have sustained in part the private enterprises' tendency to depend on debt financing rather than on equity financing. Whenever their long-term capital needs cannot be satisfied domestically, firms go to the foreign currency system to borrow. With relatively lower interest rates and forward cover usually provided by the Central Bank, or with guarantees given by the government to minimize exchange risk and/or default, it is less costly for firms to borrow from the foreign currency system than to borrow directly from domestic banks. Other disadvantages, such as the money-supply implications of OBUs lending to non-bank entities, will not be discussed in this paper.

Floating Interest-rate period, 1981–1984

Interest-rate ceilings on all types of deposits and loans (except short-term loans) were finally lifted in July 1981. Those on short-term loans were eventually lifted in January 1983. The reserve requirements of all deposit liabilities of commercial banks were supposed to be gradually scaled down from 20 per cent in 1980 to 16 per cent in 1982, with the aim of making available large amounts of funds to borrowers and, at the same time, reducing the bank intermediation cost, which would hopefully be passed on to borrowers in terms of a reduced borrowing rate.

The interest-rate reform was only part of a larger package of financial reforms introduced in 1980. The entire financial system was restructured, primarily to increase competitive conditions in the financial system with resulting greater efficiency and also to increase the flow of funds available to borrowers on medium- and long-terms. It is to be noted that the interest-rate reforms effected between 1974 and 1980 were a modest success in terms of increasing the flow of funds to long-term deposits. However, the 1972 financial reforms formally enforced financial specialization, thereby lessening direct competition among various types of financial institutions.

The new policy framework calls for a reduction in differentiation among categories of banks and non-banks authorized to perform quasi-banking (NBQBs), and for the promotion of large banks (called "expanded commercial banks" or "universal banks") offering a broader range of financial services, including those heretofore reserved for investment houses, such as underwriting and securities dealing. One further innovation here is that Unibanks are allowed to go into equity investments in both allied and non-allied undertakings, with some restrictions, of course. With this approach banks were expected to infuse fresh blood into an anaemic equity market, since big banks have sufficient resources to participate in that market. Likewise, restrictions on the operations of other financial institutions were also relaxed. For example, rural banks may engage in branch banking, and are no longer bound to lend to small farmers.

The authorized activities of various financial entities and the limits on equity investments by banks are summarized in tables 25 and 26 respectively.

The recent financial reforms stress size, as may be gathered from the minimum capital requirements for each type of financial entity shown in table 27. The objective of increasing the minimum capital is twofold: one is to enable bigger banks to exploit economies of scale, and the other to provide larger and more stable sources of funds for long-term lending. Existing financial institutions can meet the necessary capital requirement through internal capital build-up and/or merger and consolidation. As of December 1984, ten banks were given licences to operate as universal banks.

The important supporting mechanism of the new thrust to induce financial institutions to go into long-term financing is the newly opened special rediscounting window of the Central Bank. Specifically, the "medium- and long-term rediscounting" facility allows banks to rediscount papers evidencing medium- and long-term loans extended by them for the acquisition of fixed assets, working capital in connection with a proposed or ongoing expansion development programme, investment in affiliates and other institutions, and investment in high-grade securities. To encourage banks and NBQBs to engage

Table 25. Authorized activities of various financial entities based on the amended banking laws

Authorized activities	Expanded commercial banks (Unibank)	Commercial banks (KB) Domestic	Commercial banks Foreign	Savings and mortgage banks	Thrift banks/private development banks	Savings and loan association	Rural banks	Investment houses (IH)
Commercial banking services								
1. Accept deposits	1	1	1	1	1	1	1	3[a]
2. Issued LCs and accept drafts	1	1	1	1[b]	1[b]	1[b]	3[c]	3[c]
3. Discounting of promissory notes and commercial papers	1	1	1	2	2	2	2	1
4. Foreign exchange transactions	1	1	1	3	3	3	3	1
5. Lend money against security	1	1	1	1	1	1	1	1
Nationwide branching operations	1	1	1	1	1	1	1	1
Equity investment in allied undertakings[d]	1	1	1	1	1	1	1	1
Equity investment in non-allied undertakings	1	3	3	3	3	3	3	1

Table 25 (*continued*)

Authorized activities	Expanded commercial banks (Unibank)	Commercial banks (KB) Domestic	Commercial banks (KB) Foreign	Savings and mortgage banks	Thrift banks/private development banks	Savings and loan association	Rural banks	Investment houses (IH)
Trust operation	1	2	2	2	2	2	2	1
Issued real estate and chattel mortgage, buy and sell them for its own account, accept/receive in payment or as amortization of loan	1	1	1	1	1	1	1	1
Direct borrowing with Central Bank	1	1	1	1	1	1	1[e]	1
Activities of an investment house								
1. Securities underwriting	1	3	3	3	3	3	3	1
2. Syndication activities	1	3	3	3	3	3	3	1
3. Business development and project implementation	1	1	1	1	1	1	1	1

4. Financial consultancy and adviser–trust operation	1	1	1	1		2	1
5. Portfolio management–trust operation	1	1	1	1		2	1
6. Mergers and consolidation	1	1	1	1		1	1
7. Research and studies	1	1	1	1		1	1
8. Acquire, own, hold, or lease real and/or personal	1	1	1	1			1
9. Pension, profit-sharing pension benefit funds–trust operation	1	1	1	1		2	1
10. Money-market operation	1	3^f	3^f	3^f		3^g	1

1 = Authorized activities; 2 = Authorized but subjected to Monetary Board approval; 3 = Not authorized/prohibited.
a. IH are not yet allowed to accept deposits. However, certificates of time deposits for commercial banks and thrift banks have been put on an equal footing with the money-market instruments of investment houses by subjecting them both to the same tax rate of 20 per cent.
b. Limited only to domestic LCs and drafts.
c. This may be allowed for IHs that finance imported equipment but not for raw material requirements.
d. Includes warehousing companies, leasing companies, storage companies, safe deposit box companies, companies engaged in the management of mutual funds banks, and such other similar activities as the Monetary Board may declare to be appropriate.
e. As decreed in PD 1685, amending PD 1309, allowing the CB to grant portions of its foreign loans to financial institutions other than banks.
f. Full-blown money-market operation, which includes deposit substitutes, requires a quasi-banking licence.
Source: PDCP, "Universal Banking in the Philippines," *Philippine Business Review*, 13 (1980): table 1.

Table 26. 1980 financial reforms: Limits on equity investments by banks (percentages)[a]

Activities	Limits for commercial banks	Limits for universal banks[b]
Allied undertakings		
Financial allied undertakings		
Commercial banks	30	30
Thrift banks (private development banks, savings and mortgage banks, savings and loan banks, stock savings, etc.)	100	100
Rural banks	100	100
Investment houses	40	100
Others (leasing, credit card venture capital companies, etc.)	40	100
Non-financial allied undertakings		
Warehousing companies	100	100
Storage companies	100	100
Safe deposit box companies	100	100
Mutual fund management companies	100	100
Computer service companies	100	100
Insurance agencies	100	100
Home building/development companies	100	100
Agricultural drying or milling companies	100	100
Non-allied undertakings		
Agriculture	0	35
Manufacturing	0	35
Public utilities	0	35

a. Limits setting only a minority equity investment in a single enterprise can be waived upon the approval of the President.
b. A universal bank or a commercial bank with expanded functions has a minimum capitalization of 500 million pesos.
Source: CB Circular No. 739, pp. 50–57.

in term transformation, a "lender of last resort" facility was opened. Any paper, irrespective of maturity, is acceptable security for this facility, which has a maturity period of not more than 90 days. Banks and NBQBs encountering temporary liquidity problems while engaging in term transformation may avail themselves of this facility. However, to minimize moral hazards, banks are charged a rate closer to the market rate.

Table 27. Minimum capitalization of private domestic banks and non-banks authorized to perform quasi-banking activities (NBQB)

Type of institution	Minimum capitalization (million pesos)
Universal banks	500
FCDUs	150
Commercial banks	100
Thrift banks	
New thrift banks	
Metro Manila	20
Other places	10
Existing banks	
Metro Manila	10
Other places	5
Rural banks:	
New Rural Banks to be established must have 0.5 million pesos before they can operate. Existing rural banks are allowed to increase their capital within a period of time depending upon their number of years of operation.	
Non-bank quasi-banks	20
Investment houses (IH)	
New NBQBs other than IH	20

Source: CB Circular No. 739 (1980).

To prop up the stock market, a "stock financing" facility was opened. This facility provides assistance to securities dealers/stockbrokers and investors through banks in order to provide liquidity for secondary trading in high-grade stocks or blue chips.

A rationalization programme of government securities was also initiated in 1981. The Central Bank started to phase out its CBCIs in that year in order to make way for the Treasury Bills, which will become the primary government securities in the securities market. It should be noted that the CBCIs issued in the 1970s were mainly utilized to rechannel funds from the urban to the rural areas, and were therefore not effective instruments of monetary policy.

The period 1981–1984 was not a very good one for the Philippine economy in general, and for the financial system in particular. The country was just nursing the deep wounds inflicted by the 1979–1980 oil shock when a liquidity crisis struck in the early part of 1981. This brought down a number of financial institutions, especially those active in the money market, including one big investment house which had played a leading role in introducing financial innovations into the system. The crisis exposed the weaknesses of the short-term money market and led to changes in the regulatory framework.

As it started to recover, the financial system was struck again by a balance-of-payments crisis, far more severe than the 1981 liquidity crisis. This time, more banks collapsed, including the biggest savings bank in the country. Most of the painful effects were felt in 1984. A great number of firms went bankrupt, while those remaining started retrenching. It is against this background that new financial liberalization efforts were launched.

The floating interest-rate regime left the determination of the interest rates to the market. More aggressive banks offered higher rates to attract more deposits. Thus, increased competition has offered savers competitive rates for their surplus funds.

The lifting of interest-rate controls sent the rate on time deposits upwards from 10.5 per cent per annum in 1980 to 15.6 per cent in 1981 (see table 9). It declined thereafter, but sharply rose again to 32.5 per cent in 1984, when inflation was galloping at the rate of 50 per cent per annum. The rate on savings deposits, however, was only inching up, owing to the reluctance of a lot of banks, especially bigger ones, to increase the deposit rate. The intention was to give more reward to savers who keep their funds in longer-term deposits. Indeed, the share of time deposits in total bank liabilities is more pronounced during the floating rate period, although the average growth rate decelerated because of the depressed economic activity (see table 20).

Lending rates increased, but not as much as most people expected them to, except in 1984 when the inflation rate was very high and when the Central Bank increased the rates for both the CB Bills and Treasury Bills.

The impact of floating interest rates on the flow of loanable funds can be clearly discerned from table 9. The M3/GNP ratio consistently rose from 25.6 per cent in 1980 to 29.8 per cent in 1983, the highest ever achieved by the Philippines. The interest rates on time and savings deposits were substantially positive during this period. However, the ratio dropped precipitously in 1984 when the real interest rates on both traditional deposits and deposit substitutes were severely negative. The collapse of several banks and the high inflation rate in 1984 led to substantial disintermediation. This is fundamentally similar to what happened during the last two years of the transition period.

Still, the performance of the Philippines in terms of increasing the flow of loanable funds during the floating interest-rate period cannot compare with that of some other Asian countries.

The floating interest-rate policy yielded favourable results in terms of lengthening the term structure of banks' loan portfolio. Table 13 shows that medium- and long-term loans outstanding had been increasing both in absolute and relative terms during the period 1981–1984. The lengthening in the term structure of their deposit liabilities and the floating interest-rate policy, which allowed banks to charge clients the market rate of interest, could have provided banks with incentives to allocate more funds for medium- and long-term financing. However, the amount of long-term financing currently provided by banks still leaves much to be desired, considering the investment requirements of the country.

The equities market remained sluggish during the period 1981–1984. It was not, therefore, a significant source of medium- and long-term finance. Firms continued to depend on debt financing, as may be observed from the steadily rising, average debt–equity ratio for all industrial sectors (see table 15).

The Nomura Research Institute study[18] attributed the sluggish condition of the equities market to the low supply of and demand for equity shares. The low supply could have been caused by the following factors: (1) inadequate tax incentives; (2) the high cost of obtaining capital from the equity market, as demonstrated by the study's empirical analysis; (3) the reluctance of family-held corporations to go public; and (4) the very limited number of growing enterprises. On the other hand, the low demand for securities be could be attributed to the following factors: (1) the equity investment was less attractive than other investment instruments, like money-market instruments, in terms of both return and liquidity; (2) there was general discontent with the equities market, especially with the recently uncovered malpractices and fraudulent tradings perpetrated by a few participants; and (3) securities brokers/dealers lack the necessary resources to support greater trading volumes.

The high dependency of firms on debt financing rather than on equity financing during the floating interest-rate period when lending rates were inching up may have been sustained by the Central Bank's liberal rediscounting policy. Between 1980 and 1983, a lot of rediscounting windows had been opened with prescribed maximum lending rates significantly below the market rates (see table 28). Indeed, the Central Bank's selective credit programme lost its selectiveness, since many economic activities could qualify for rediscounting. The relatively attractive spread between the discount and prescribed loan rates provided private banks with incentives to secure more funds from the Central Bank, instead of significantly increasing their intermediation activities. Indeed, when the Central Bank tightened up its rediscounting windows in 1984 in view of the economic crisis, many banks found themselves illiquid.

Between 1981 and 1984, commercial banks were able to borrow from the Central Bank, through its rediscounting windows, the amount of 209,115 million pesos, which was more than twice the amount that they availed of in the past ten years (see table 29). This could be one of the reasons why the M3/GNP did not increase dramatically during the floating interest-rate period. Moreover, specialized government financial institutions intensified their borrowings from the Central Bank during the same period, providing low interest credits to certain sectors.

Table 28. Rediscounting windows of the Central Bank

Facility	Implementing circular	Date	Loan value (%)	Rediscount rate (%)	Lending rate (%)	Maturities
Regular rediscounting						
1. Supervised credits	784	27 Feb. 1981	100	3	12	120 days
2. Non-supervised credits	784	27 Feb. 1981	80	8	14	60/120/270 days
3. Small/medium-scale industry	784	27 Feb. 1981	80	8	14	120 to 270 days
4. Exports						
Non-traditional	784	27 Feb. 1981	80	3	12	90 days
Traditional	784	27 Feb. 1981	80	8	14	10–40 days for sight drafts/120–170 days for production credits
5. Masaganang Maisan	828	9 Oct. 1981	100	3	15	120–270 days
6. Special programmes						
NGA, FTI	784	27 Feb. 1981	100	3	6	180 days
Grains Quedan/ food Quedan	881	25 June 1982	80	3	10	190 days
7. Tax credit certificates	802	1 June 1981	80	8	14	180 days
8. Tobacco trading	715–801	1 Feb. 1980/ 1 June 1981	80	8	14	180 days
						One to five years; or the maturity of the paper/last amortization, whichever comes first

FINANCIAL LIBERALIZATION: THE PHILIPPINE CASE   255

9. Energy-generating projects	803–872	18 June 1981/ 26 Apr. 1982			
Mini-hydro			100	3	10
Dendro thermal			100	3	8
10. Stock financing	807–851	26 June 1981/ 15 Feb. 1982	80	8	14	180 days
11. Metal financing	873	6 May 1982	80	8	14	90 days with another 90 days roll-over
12. Dollar rediscounting	875–944	21 May 1982/ 15 Aug. 1983				90 days
Dollars sold to CB			100	12	—	90 days, renewable for 90 days
Dollars deposited with CB			100	12	—	360 days or maturity, whichever comes first
13. Manpower exporters	842–894–895	29 Jan. 1983/ 24 Sept. 1982	80	3	12	180 days
14. Orchard growing/ upland farming	Circular letter	23 Oct. 1982	80	8	14	120 days
15. Congress organizers	910	6 Jan. 1983	80	3	12	180 days
16. Coconut millers/ sicators	921	28 Mar. 1983	80	3	12	90 days

Table 28 (continued)

Facility	Implementing circular	Date	Loan value (%)	Rediscount rate (%)	Lending rate (%)	Maturities
Special rediscounting						
1. Medium- and long-term	846	1 Feb. 1982				
– For acquisition of fixed assets			75	11	15	Up to 10 years non-renewable
– For working capital in connection with a proposed or ongoing expansion development programme			75	11	15	Up to 3 years non-renewable
– For investment in affiliates and other institutions			70	14	—	Up to 7 years non-renewable
– For investment in high-grade securities			70	14	—	Not to exceed one year

FINANCIAL LIBERALIZATION: THE PHILIPPINE CASE 257

2. Lender of last resort	749–862–864–907	Aug. 1980 23 Mar. 1982 24 Dec. 1982	80% or as may be provided for under an MB resolution	
Commercial banks			MRR plus 2% or more	90 days
Thrift banks			MRR plus 2% or more with 5% liquidity	60 days
NQBQs			24–32% for loans ranging from P150M–300M & over; plus 2% for each roll-over	
Emergency rediscounting	Sec. 90, RA 265/907	24 Dec. 1982	16% or MRR plus 3%	90 days
Rediscounting with the government	Sec. 95, RA 265			

Source: San Jose (1983), pp. 16–17.

Table 29. Total Central Bank loans granted through rediscounting by type of institution (in million pesos)

Year	Total	Public sector[a]	Commercial banks	Other banks[b]	Non-banks
1970	2,069	414	1,413	242	—
	(100.0)[c]	(20.0)	(68.3)	(11.7)	
1971	1,179	327	359	493	—
	(100.0)	(27.7)	(30.4)	(41.8)	
1972	1,188	449	276	493	—
	(100.0)	(37.8)	(23.2)	(40.0)	
1973	1,330	15	779	536	—
	(100.0)	(1.1)	(58.6)	(40.3)	
1974	4,491	214	3,222	1,055	—
	(100.0)	(4.8)	(71.7)	(23.5)	
1975	8,583	188	6,562	1,833	—
	(100.0)	(2.2)	(76.4)	(21.4)	
1976	10,729	300	9,006	1,423	—
	(100.0)	(2.8)	(83.9)	(13.3)	
1977	9,457	817	7,067	1,573	—
	(100.0)	(8.6)	(74.7)	(16.6)	
1978	15,513	2,342	11,077	2,094	—
	(100.0)	(15.1)	(71.4)	(13.5)	
1979	29,538	2,016	24,197	3,325	—
	(100.0)	(6.8)	(81.9)	(11.2)	
1980	43,914	2,535	37,442	3,937	—
	(100.0)	(5.8)	(85.3)	(9.0)	
1981	54,884	2,516	43,762	6,546	2,060
	(100.0)	(4.6)	(79.7)	(11.9)	(3.8)
1982	46,051	2,516	36,659	6,876	—
	(100.0)	(5.5)	(79.6)	(14.9)	
1983	40,132	6,034	27,202	6,322	574
	(100.0)	(15.0)	(67.8)	(15.8)	(1.4)
1984[d]	24,138	7,017	14,288	2,590	243
	(100.0)	(29.1)	(59.2)	(10.7)	(1.0)

a. Public sector: National, local, and semi-government.
b. Other banks: Specialized banks, rural banks, and thrift banks.
c. Figures in parentheses are percentages.
d. Preliminary.
Source: Department of Economic Research, Central Bank.

Concluding Remarks

This paper has reviewed the financial policies of the country that led to financial repression and the subsequent efforts to liberalize the financial system. The liberalization efforts focused on changes in interest-rate policy and on the structure of the financial system, with the aim of developing the capital market, especially the long-term capital market.

The financial policies of the 1950s laid the groundwork for a repressed financial system. Specifically, the low interest-rate policy, rediscounting policy, and some regulations that led to a higher degree of financial fragmentation restrained the growth of the financial system and, at the same time, paved the way for the emergence of family-held corporations highly dependent on debt financing.

The equities market has long been an insignificant source of medium- and long-term capital. Financial policies are partly to blame for this. So far, recent financial liberalization efforts have not altered this situation at all. In the absence of a smoothly functioning equities market, the pressure to provide long-term funds for investment falls heavily on banks. In fact, the Philippine capital market up to the present is still dominated by the banking system. Increasing the flow of loanable funds is, therefore, a necessary condition if banks are to fulfil this responsibility.

The financial liberalization efforts, especially the most recent ones, have basically addressed this issue. Although modest successes have been achieved in certain areas, for example the slight increase in the proportion of medium- and long-term deposits and loans, in general the results still leave much to be desired. In table 30 we reproduce some of the variables presented above in order to underline certain important points.

One can see immediately from the table that as the economy moved from the rigid and low interest-rate regime to the transition period, the average M3/GNP ratio increased from 23 to 27 per cent, indicating that the move towards a regime in which interest rates were still administratively fixed but constantly adjusted by the Monetary Board to reflect market conditions did produce favourable results. However, the move towards a fully flexible interest-rate regime did not increase at all the flow of loanable funds.

Why did the switch to a fully liberalized regime fail to produce the desired results? To answer this question, we have to look at other aspects of the operating financial policy framework. During the period of financial liberalization, the Central Bank opened a lot of rediscounting windows, which all but nullified the selectiveness of its selective credit programme. This cause disintermediation in two ways. First, both private and government-owned banks took this opportunity of increasing their borrowings from the Central Bank. Thus they became mere conduits of Central Bank funds instead of intermediaries mobilizing funds of surplus units. Second, the inflationary impact of this expansionary policy discouraged surplus units from investing their funds in financial assets yielding negative real returns. The table shows that the floating interest-rate period was characterized by an average inflation rate substantially higher than that of the previous regimes.

Table 30. Selected financial and economic indicators (figures are averages for the period)

Regime	Financial ratios				Real interest rates		Lending rate[a]	GNP growth rate	Inflation rate	Budget deficits[b]
	M2/M1	M3/M2	M2/GNP	M3/GNP	SD	TD				
Low interest-rate regime, 1956–1973	1.80	1.02	0.23	0.23	−1.84	−1.06	5.69	5.28	6.31	0.56[c]
Transition period, 1974–1978	2.18	1.35	0.20	0.27	−7.21	−3.78	−1.93	6.14	14.5	1.22
Floating interest-rate period, 1981–1984	3.06	1.20	0.22	0.27	−10.58	−1.62	2.6	0.19	20.78	2.89

a. For less than 730 days' secured loans.
b. As a percentage of GNP.
c. From 1966 to 1973.

Another aspect that must be examined is the fiscal deficit. As is clear from the table, the average fiscal deficit for each subperiod has been rising and seems to be positively correlated with the average inflation rates. It should be noted that a significant portion of the budget deficit was financed by the Central Bank. For example, borrowings by government from the Central Bank through the "rediscounting with the government" facility averaged 2.5 billion pesos annually between 1980 and 1982. It had already reached 3.5 billion pesos by the first half of 1983.[19] In order to help bring down inflation, therefore, fiscal discipline must be instilled; at the very least, the Central Bank should not be called upon to finance budget deficits. However, this is a tall order, given the fact that the Central Bank is not entirely independent of the fiscal sector. Perhaps it is time that it were made so, since then the fiscal sector would be forced to be extremely careful in managing its deficits, while the Central Bank could pursue its task of stabilizing the economy more effectively.

Finally, we should take note that the period 1981–1984 was marked by a large number of bank failures. The relative absence of regulatory control by the Central Bank over bank portfolios allowed banks to take on more risky assets. Interestingly, the recent collapse of many banks can be traced to questionable loans made by banks to their directors, officers, stockholders, and related interest (DOSRI), which were not closely monitored by the Central Bank. This suggests that even if the financial system is deregulated, certain forms of regulation must be exercised by the Central Bank to reduce riskiness in banks' portfolios – something that is essential if people's confidence in the system is to be restored. Perhaps standard financial ratios or other financial ratios modified to suit the characteristics of Philippine financial institutions could be adopted to ensure their stability as the financial system moves from a repressed to a liberalized one.

Acknowledgements

The author is grateful to Dr. Baldomero Regidor for his valuable comments and suggestions, to Ms. Vangie Yu for her able research assistance, and to Ms. Emma Pizaro-Cinco for painstakingly typing the manuscript.

Notes

1. A good description of the capital market can be found in Nomura Research Institute, "A Capital Market Study of the Philippines" (ADB Development Policy Office, 1984; and IMF/WB Joint Mission Report, "Aspects of the Financial Sector: The Philippines" (1980).
2. A comprehensive description of the money market can be found in Victoria S.

Licuanan, "An Analysis of the Institutional Framework of the Philippine Short-term Financial Markets" (Philippine Institute for Development Studies, Manila, 1986).
3. Jaime C. Laya, *A Crisis of Confidence and Other Papers* (Central Bank of the Philippines, Manila, 1982).
4. See note 3 above, p. 13.
5. Edita A. Tan, "Philippine Monetary Policy and Aspects of the Financial Market: A Review of the Literature," in PIDS, *Survey of Philippine Development Research I* (NEDA–APO Production Unit, Manila, 1980), p. 194.
6. The M3/M2 ratio is irrelevant at this point since data on deposit substitutes became available only in 1973.
7. Rosa Linda G. Valenzona, "Long-term Capital Market for Firms' Investment Financing," unpublished M.A. thesis (UP School of Economics, 1969).
8. See note 7 above.
9. See note 2 above.
10. World Bank, *The Philippines: Priorities and Prospects for Development* (NEDA Production Unit, Manila, 1977), p. 357.
11. See note 3 above, p. 162.
12. FCDUs were first established in 1970. However, their performance prior to 1936 was unsatisfactory in view of the restrictive policies governing their operations.
13. Maria Socorro L. Africa, "Participation of Offshore Banking Units (OBUs) in the Peso Market: Implications for Domestic Credit and the Money Supply," unpublished M.A. thesis (UP School of Economics, 1980), p. 26.
14. 343 banks are banks that were granted the privilege of accepting foreign currency deposits under CB Circular, no. 343, 21 July 1970.
15. See IMF/WB Joint Mission Report (note 1 above).
16. See note 2 above.
17. A good discussion of the role of OBUs in the domestic capital market can be found in M.S.L. Africa (note 13 above).
18. See note 1 above.
19. Armida S. San Jose, "Central Bank Rediscounting Operations," *CB Review*, September 1983.

Bibliography

Africa, Maria Socorro L. "Participation of Offshore Banking Units (OBUs) in the Peso Market: Implications for Domestic Credit and the Money Supply." Unpublished M.A. thesis. University of the Philippines School of Economics, Diliman, Quezon City, 1980.

Central Bank of the Philippines. *Central Bank Statistical Bulletin*. Various issues.

IMF/World Bank Joint Mission Report. "Aspects of the Financial Sector: The Philippines." Washington, D.C., 1980.

Kim Wan-Soon. "Financial Development and Household Savings: Issues in Domestic Resource Mobilization in Asian Developing Countries." ADB Economic Staff Papers, no. 10. ADB, Manila, 1982.

Laya, Jaime C. *A Crisis of Confidence and Other Papers.* Central Bank of the Philippines, Manila, 1982.

Licuanan, Victoria S. "An Analysis of the Institutional Framework of the Philippine Short-term Financial Markets." Philippine Institute for Development Studies, Manila, 1986.

McKinnon, Ronald I. "Financial Repression and Liberalization Problems within Less Developed Countries." In: Sven Grassman and Erik Lundberg, eds., *The World Economic Order, Past and Prospects.* Macmillan, Hong Kong, 1981.

———. "The Order of Economic Liberalization: Lessons from Chile and Argentina." In: K. Brunner and A. Meltzer, eds., *Economic Policy in a World of Change.* North-Holland, Amsterdam, 1982.

Nomura Research Institute. "A Capital Market Study of the Philippines." ADB Development Policy Office. ADB, Manila, 1984.

Patrick, Hugh, and Honorata A. Moreno. "The Evolving Structure of the Philippine Private Domestic Commercial Banking System from Independence to 1980: In Light of Japanese Historical Experience." UPSE seminar paper. University of the Philippines School of Economics, Diliman, Quezon City, 1982.

Report of the Inter-Agency Committee on the Study of Interest Rates. National Economic Council, Manila, 1971.

San Jose, Armida S. "Central Bank Rediscounting Operations." *CB Review.* Central Bank of the Philippines, Manila, September 1983.

Suleik, Mercedes B. *Towards Market-determined Interest Rates: The Philippine Experience.* Staff Papers, no. 8. Central Bank of the Philippines, Manila, 1983.

Tan, Edita A. "Philippine Monetary Policy and Aspects of the Financial Market: A Review of the Literature." In: PIDS, *Survey of Philippine Development Research I.* NEDA–APO Production Unit, Manila, 1980.

———. "The Structure and Growth of the Philippine Financial Market and the Behavior of Its Major Components." PIDS Working Paper, no. 81-06. PIDS, Manila, 1981.

———. "Development Finance and State Banking: A Survey of Experience." PIDS Staff Paper Series, no. 84-04. PIDS, Manila, 1984.

Valenzona, Rosa Linda G. "Long-term Capital Market for Firms' Investment Financing." Unpublished M.A. thesis. University of the Philippines School of Economics, 1969.

Villegas, Edberto M. *Studies in Philippine Political Economy.* Silangan, Manila, 1983.

World Bank. *The Philippines: Priorities and Prospects for Development.* NEDA-APO Production Unit, Manila, 1977.

Appendix

Type of national government issues	Term	Price	Interest rates (p.a.)	Manner of sale	Taxability	Eligibility and other features
Direct obligations of the national government						
1. Treasury Bills						
Auctioned (weekly)	15, 35, 49, 63, 91, and 182 days	Discount basis	Dictated by the result of auction	Auction	Taxable but exempt from withholding tax	Eligible as security in any transaction with the government
Negotiated	Dictated by the results of the negotiation	Discount basis	Depending on the result of the negotiation	Negotiation or on tap	As above	As above
2. Treasury Notes						
Marketable Auctioned	5 yrs	Yield basis (at a discount)	9% p.a. payable semi-annually	Auction/ negotiation or on tap	As above	1. Agri/Agra eligible 2. Eligible as collateral
Regular	10–15 yrs	Par value	8.5–11.75%	Over the counter	As above	1. See note a 2. Callable in whole or in part 3. Eligible as reserve for life and non-life insurance companies

FINANCIAL LIBERALIZATION: THE PHILIPPINE CASE   265

Non-marketable	2–25 yrs	As above	2–7%	As above	As above	1. Eligible as collateral 2. See note b

3. Treasury Bonds

Marketable	Above 5 yrs	Par value	10.75	Over the counter	As above	1. Callable in whole or in part 2. See note b 3. Eligible as collateral 4. Eligible as reserve for life and non-life insurance companies
Non-marketable	As above	As above	2–7%, payable semi-annually	As above	As above	See note b

Guaranteed obligations of the national government

1. National Development Company Bonds	5 yrs (later issues were reduced to 2 yrs)	Yield basis (at a discount)	Coupon rate of 9% payable semi-annually	Auctioned and/or negotiated	As above	1. Agri/Agra (1st to 3rd series only) 2. Eligibles as reserve for life and non-life insurance companies
2. Light Rail Transit Notes	3 yrs	As above	As above	As above	As above	May be callable in whole or in part before maturity

Appendix (*continued*)

Type of national government issues	Term	Price	Interest rates (p.a.)	Manner of sale	Taxability	Eligibility and other features
3. Public Estates Authority Bonds	15 yrs	Par value	8.5–14% (payable semi-annually)	Negotiated	Tax-free/taxable[c]	
4. National Housing Authority Bonds	5 yrs	Yield basis (at a discount)	coupon rates (8.5 to 9%)	Auction or direct disposition	1. First issue tax-exempt 2. Second issue taxable	1. Callable in whole or in part 2. Considered as authorized investments of insurance companies
Government corporate issues not guaranteed by the national government						
1. Philippine Charity Sweepstakers Bonds	10 yrs	Par value	8.5%, payable semi-annually[d]	Auction/negotiated	Tax-exempt	To be secured by real estate mortgage with a security coverage of 167% of outstanding principal balance

FINANCIAL LIBERALIZATION: THE PHILIPPINE CASE

Philippine government securities denominated in foreign currencies

1. US dollar denominated Treasury Bills	91-day, 182-day, 364-day series	Discount basis	At rates acceptable to the MOF with LIBOR as reference	See note e	Payable on or after maturity in US dollars
				As above	

a. Those issued prior to circular 638 dated 6 November 1978 are eligible as reserves against deposit substitute liabilities.
b. Those with interest rates up to 4 per cent are eligible as reserves against deposit and deposit substitute liabilities. Those exceeding 4 per cent but issued prior to the promulgation of circular 638 are eligible as reserves against deposit substitute liabilities.
c. First issues, carrying 8.5 per cent interest, are tax-free; later issues, carrying 14 per cent interest, are taxable.
d. Net of transaction, tax thereon, which will be paid by PESO (Philippine Charity Sweepstakers Office).
e. Phase 1: negotiated (with one or two local dealers); phase 2: negotiated (open to one or two local dealers and international dealers); phase 3: auction (open to local and international dealers).

Source: Nomura Research Institute (1984).

THE FINANCIAL SYSTEM AND DEVELOPMENT: THE FORMATION OF FINANCIAL INSTITUTIONS FOR LABOUR-INTENSIVE SECTORS

Akio Hosono

As is well known, the economic development of Japan has been accomplished fundamentally within the framework of a free-market system. The commercial banks have been the principal mechanism by which the financial resources for investment and working capital of Japanese enterprises have been mobilized.

Nevertheless, because of the fact that Japan was a latecomer to modern economic growth, and therefore passed quite recently through the phase of intermediate development (almost equivalent to the phase in which the contemporary NICs find themselves), there were sectors which, in spite of their importance from a macro-economic point of view, were not duly catered to by commercial banks within the mechanism of a free capital market.

In this paper, the author will discuss: (1) the characteristics of such sectors and the reasons why they were not appropriately serviced by commercial banks; (2) general aspects of the financial system in Japan that were created and strengthened to support such sectors; and (3) the formation of financial institutions for labour-intensive sectors; finally (4) I shall make a few observations regarding the impact of the liberalization of the capital market on the financial system for small and medium-sized enterprises.

I should like to give special emphasis to point 3, since this has great relevance for today's developing countries in terms of their promotion of "growth and equity."

Some Basic Issues concerning the Assignment of Financial Resources to Priority Sectors at the Intermediate Phase of Development

It goes without saying that under the free-market mechanism financial resources are normally available to sectors or enterprises that are characterized by a high rate of profit and overall creditworthiness. We must of course recognize that it is the highest-priority sectors or enterprises that naturally receive the largest amount of financial resources, specially at an advanced phase of industralization.

Nevertheless, there are at least two important areas for economic development, especially at the intermediate development phase, which do not normally receive the necessary finances through the free-market mechanisms.[1] These areas are related to the following development goals: (1) the promotion of new industries and (2) support for small units of production based on labour-intensive methods with high capacity for labour absorption. Their access to the capital market is limited because of their low rate of profit in the short or medium term and/or their lack of creditworthiness.

As regards the first area, we would cite industries which lack a comparative advantage in the world market, but could be competitive in the future. In other words, they are those industries that have a comparative advantage from a dynamic point of view, but which, generally speaking, do not receive adequate financial support because they cannot guarantee a high rate of profit in the short term.

Concerning this aspect, it is worth while citing the following paragraph from the *White Paper on the Japanese Economy 1984*, which summarizes the structural changes in Japan's trade and industry over the last two decades:

Our country, characterized by a shortage of natural resources, adopted the policy of improving the competitiveness of those industries which had a dynamic comparative advantage, and which depended on imported primary products in order to develop economically in an open world system. Thus, Japan consolidated its advanced industrial structure by incessant efforts in technological innovation and intensive capital formation. However, with regard to those industries that were becoming uncompetitive because of their intensive use of natural resources and/or labour, efforts were made to adapt them through active industrial adjustment and the expansion of investment in these industries overseas. As a result of these policies there has been a structural transformation, which can be summarized as follows: The leading industries of Japan in the early period were labour-intensive – for example, textile industries. These were replaced by new leading industries that made intensive use of capital and natural resources, such as metallurgy and chemical industries. These have been superseded by technology-intensive industries such as transport equipment and electrical and non-electrical machinery.

Certain industries that are likely to have a comparative advantage in the future are unable to attract sufficient financial resources from the capital market owing to, among others, the following factors: (1) commercial banks do not have enough information concerning the trends in the dynamic international division of labour; (2) investment in such industries takes a long time to mature; (3) even in cases in which the banks have sufficient information, there still remains a degree of uncertainty.

In order that these industries may obtain the necessary resources, the introduction of certain measures to complement the mechanisms of the free capital market is indispensable. With regard to the first factor, it is important for the government to put forward plans, based on thorough analysis, concerning the future industrial structure, as well as policy measures to support industries that have a dynamic comparative advantage, in order to convince the commercial banks to finance such industries. In this sense, the support of the government or other institutions (including international financial institutions) for these industries is crucial, even if the amounts involved are small, because the fund could work as a catalyst.

As for the second factor, reinforcement of the long-term credit system is particularly important. If in the free capital market a sufficient amount of long-term finance is not available, it is necessary to promote the formation of financial institutions that can mobilize long-term credit.

As for the third factor, in some cases it may be necessary to introduce measures to absorb the risk of the new industries, venture business, etc.

Regarding the small units of production, although there is not enough space here to discuss fully their importance for economic development, the following three interrelated points should be stressed:

1. As Ohkawa, Tajima, and Motai observed with regard to Japanese manufacturing industry, as well as some developing countries' industries, and Hosono in discussing some Latin American examples, small and medium-sized enterprises, both at the aggregate level of manufacturing industry and at the disaggregate two-digit level of industrial classification, are characterized by a higher output capital ratio (higher productivity of capital – that is, higher effciency of capital utilization) than large enterprises.[2] This fact has major significance for developing countries because their major constraint to economic development is scarcity of capital.
2. As is well known, these industries are important in absorbing the labour force, because they normally use labour-intensive methods of production.
3. In addition to these aspects, these industries could make an important contribution to the expansion and diversification of exports. In those developing countries with an abundant supply of cheap labour, a comparative advantage is found in the products of the labour-intensive sectors, particularly in those of small and medium-sized enterprises. As the level of real wages of unskilled labour is fixed, at least until the so-called turning-point, any in-

crease in labour productivity is reflected in increased competitiveness or investment.

Nevertheless, in spite of their importance, small and medium-sized enterprises receive insufficient financial resources because of, among other things, the following factors:
1. Their creditworthiness is low, owing to the small scale of production, limited assets, and the uncertainties associated with the competition between them.
2. The cost of small-scale financial intermediation is normally higher than that of large-scale intermediation.

As in the first area, it is necessary to introduce certain measures to solve these problems in order to mobilize funds for small and medium-sized enterprises. This is one of the important areas where some arrangements need to be made to complement the functioning of the free capital market.

With respect to factor 1, some measures to absorb the risk of credit to these enterprises should be adopted. As to factor 2, some organization (such as an association or grouping of these enterprises) could enlarge the scope of financial intermediation.

Here we should add also the importance of family farms for their labour-intensive agricultural methods. They are not necessarily characterized by all three previously mentioned aspects of small and medium-sized enterprises, but they are handicapped by the above two factors, and therefore their access to the formal capital market is normally limited.

Finally, it is important to recognize that the policy of supporting small and medium-sized units of production should not be considered as a merely passive measure, as though it were simply part of a social policy. On the contrary, it should be conceived of as part of the integral economic and industrial policy, and should be designed to solve the problems which these units are obliged to face owing to the imperfections of the market in capital, labour, goods, and information.[3] Although there is no space here to discuss these imperfections (we will discuss the capital market later), we would stress the importance of small and medium-sized units of production not only because of the three aspects mentioned above (high efficiency of utilization of capital, employment effect, and contribution to exports), but also because they contribute to the maintainance of competitiveness in the internal market. There are many fields where the participation of small and medium-sized industries in the market frustrates the formation of an oligopolistic or monopolistic market by a small number of large industries. In this context we should add that although a trade liberalization policy was adopted in Japan from the early 1960s on, strong internal competition constituted an important incentive to the efficiency of enterprises even before the implementation of such a policy, so that the foreign competition introduced afterwards is not always a necessary condition for the strengthening of efficiency or competitiveness.

The Role of the Government Financial System in the Promotion of High-priority Sectors in Japan

The two areas discussed in the first section (new industries with a high dynamic comparative advantage and small and medium-sized industries) have been considered areas of high priority in the economic development policy of the Japanese government and have received a high percentage of government funds. The allocation of funds has been made through the Japan Development Bank, the People's Finance Corporation, the Small Business Finance Corporation, etc.

Here we will outline briefly the general aspects of the government financial system for the promotion of high-priority sectors and for specific policy purposes, especially the system's importance in the mobilization of financial resources in Japan and the major trends in the allocation of government funds.

The major part of government funds is allocated according to the Treasury Investments and Loans Programme (*zaisei toyushi keikaku*). This is in a sense the integrated programme of most of the government's investment and loan activities, but does not include investment and expenditure from the general budget account (an important part of this budget is also used for public investment), investment from special accounts (except the Special Account for Industrial Investment), or short-term loans of the Government Trust Fund Bureau.

The Treasury Investments and Loans Programme has the following important characteristics.

1. Funds are not used for the purchase of goods and services or for transfer expenditure, and therefore they are separated from the government budgets general account.
2. The major source of funds is not tax revenue, as in the case of the general account, but private saving, and therefore loans from the Treasury Investments and Loans are naturally subject to principal and interest payment.
3. Funds are allocated to areas of public interest and of high priority from the point of view of economic policy, which are normally not covered by the private financial institutions.

Therefore, the financial activities of the Treasury Investments and Loans Programme complement those of the capital market, in which the allocation of funds is made through the free-market mechanism.

Nevertheless, the programme has been very important in Japan, not only because of its scale but also owing to its effect on the promotion of priority sectors, especially in the critical period of their development.

The major sources of the programme's funds are the Trust Fund Bureau, the Postal Life Insurance and Postal Annuity Fund, bonds issued by public corporations guaranteed by the government, loans by these corporations – also guaranteed by the government – and funds from the Special Account for In-

Table 1. Japan: Structure of treasury investments and loans programme according to major sources of funds (in 100 million yen)

	Total (A)	Special account of industrial investment	Trust Fund Bureau				Postal life insurance and postal annuity	Bonds and debentures with guarantee of government	Gross National Product (B)	A/B (%)
			Total	Postal savings	Welfare annuity	People's annuity				
1975	114,010	655	98,575	50,501	21,352	0	10,141	4,639	1,517,970	7.5
1980	218,036	167	185,316	94,869	41,435	5,169	16,887	15,666	2,408,470	9.1
1984	247,066	48	189,052	69,000	38,500	—	25,866	32,100	2,960,000	8.3
Composition (%)	100	0.0	76.5	27.9	15.6	—	10.5	13.0	—	—

Source: Economic Statistics Annual (Toyo Keizai, Tokyo, 1983 and 1984) (in Japanese).

Table 2. Japan: Principal funds and investments of financial institutions (deposits, savings, certificates of deposit, and bank debentures) (in 100 million yen)

	All banks[a]			Sogo banks[b]	Shinkin banks[c]	Shoko Chukin bank[d]		Norinchukin bank[e]	
	Banking accounts								
End of year or month	Deposits and savings	Bank debentures issued[f]	Trust accounts[g]	Deposits and savings[h]	Deposits and savings	Deposits and savings	Bank debentures issued[f]	Deposits and savings	Bank debentures issued
1955	37,243	2,816	3,178	4,118	2,948	138	423	1,262	358
1960	88,722	8,800	14,373	10,843	9,579	414	1,282	2,280	481
1965	206,531	23,069	35,353	32,200	31,138	1,393	3,394	10,075	2,425
1970	413,088	47,666	70,193	63,664	77,395	3,297	8,438	17,107	5,867
1975	929,213	117,705	181,041	159,408	195,568	9,309	24,640	38,512	13,343
1980	1,529,783	196,083	359,207	273,653	344,745	12,866	42,297	85,510	24,172
1983.12	1,948,955	262,651	407,039	347,540	437,034	16,744	58,593	139,086	31,124

a. Member Banks of the Federation of Bankers Association of Japan (city banks, regional banks, trust banks, and long-term credit banks).
b. Mutual loan and savings banks.
c. Commercial and industrial credit associations.
d. Central Bank for Commercial and Industrial Credit Associations.
e. Central Bank for Agricultural Co-operatives.
f. Since 1968, total of bank debentures issued and payments received in advance or margin for debentures. Up to 1967, as for "All Banks," includes matured debentures unpaid; as for the Shoko Chukin Bank and the Norinchukin Bank were only bank debentures issued.
g. Total of money in trust, securities investment trust (since 1955, but excluding trusts considered as investment trust since 1970), loan trusts (since 1955), pension trusts (since 1962), employees' property formation benefit trusts (since 1975), and employees' property formation investment fund trusts (since 1978). Since September 1981, outstanding total of net money in trust, pension trusts, employees' property formation benefit trusts, and loan trusts.

dustrial Investment. Of all these sources, funds from the Trust Fund Bureau are most important and consist of the postal savings and deposits of Welfare Annuity and People's Annuity systems.

The importance of the Treasury Investments and Loans Programme will be appreciated when the scale of its funding is compared with different indicators of the Japanese economy. For example, the amount of funds mobilized by the programme corresponded to 7.5 per cent of the Gross National Product of Japan in the fiscal year 1975, 9.1 per cent in 1980 and 8.3 per cent in 1984 (see table 1). On the other hand, the sum of the increases in postal savings, the deposits of the Welfare Annuity and People's Annuity systems, and bonds issued with the guarantee of the government – the programme's major sources – constituted approximately 20 per cent of the total increase in individual financial assets in Japan in the decade 1955–1965, 30 per cent in the following decade and 40 per cent in the period after 1975. It is important also to note that of the total amount of funds and investments of financial institutions (including deposits and savings in bank accounts, bank debentures issued, trust accounts,

Agricultural co-operatives	Life insurance companies[i]	Non-life insurance companies[i]	Trust Fund Bureau[j]	Postal life insurance and postal annuity[k]	Total including other financial institutions[l]	Net total, excluding overlapping accounts[m]
3,856	1,728	810	8,428	2,441	72,823	63,259
8,232	6,908	1,785	19,302	7,017	188,194	167,091
24,326	20,943	3,976	47,050	12,044	484,345	424,527
61,866	55,227	11,251	137,102	24,109	1,076,727	957,270
155,404	122,407	31,805	404,847	63,700	2,652,430	2,372,950
275,122	249,041	59,838	970,234	147,013	4,992,997	4,477,213
349,978	375,926	81,888	1,372,456	222,222	—	—

h. Instalments and deposits.
i. Total operating funds.
j. Total of source of funds, excluding reserve funds and suspense account.
k. Total operating funds.
l. Others represent credit federations of agricultural co-operatives, the Zenshinren Bank, credit co-operatives, and credit federations of fishery co-operatives; added later were the National Federation of Credit Co-operatives, National Federation of Labour Credit Associations (1962), and the Mutual Insurance Federation of Agricultural Co-operatives (1967).
m. Overlapping accounts represent total of deposits with other financial institutions (excluding deposits with the Bank of Japan), money in trust, and beneficiary certificates and bank debentures held by financial institutions; added later were certificates of deposit with others (1979).

Source: Bank of Japan, *Economic Statistics Annual*, 1983, 1984.

total operating funds of insurance companies, deposits and savings in agricultural cooperatives, funds of Trust Fund Bureau, etc.), 23.1 per cent corresponded to the funds of the Trust Fund Bureau at the end of September 1983. The corresponding figures of the end of 1955 and 1970 were 13.3 per cent and 14.4 per cent respectively (see table 2).

Now, by analysing the distribution of funds of the Treasury Investments and Loans Programme we are able to observe how the financing of priority sectors that were not serviced by commercial banks was achieved.

With respect to the first area mentioned in the first section of this paper, that is, the promotion of priority industries, especially those with a dynamic comparative advantage, the programme's funds were channelled through the Japan Development Bank. According to the finance policy of this bank, the highest-priority sectors were electricity and maritime transport, and the second-highest were coal-mining, steel, fertilizers, and machinery in the first decade of the post-war period (up to 1945–1955). Twenty-five per cent of the Treasury Investments and Loans Programme funds were channelled to electricity, maritime

Table 3. Japan: Structure of treasury investments and loans programme, by sectors supported (percentages)

	1953–55	1956–60	1961–65	1966–70	1971–75	1976–81
Support of strategic industries	23.6	16.6	9.9	6.3	3.7	2.9
Trade and economic co-operation	2.8	4.3	7.9	10.4	8.8	6.4
Regional development	5.7	9.0	7.5	4.6	3.7	2.6
Industrial infrastructure	26.4	21.6	26.1	24.3	23.2	18.1
Modernization of low-productivity sectors	18.6	20.9 (13.7)[a]	19.0 (12.9)	20.1 (15.6)	19.6 (15.2)	22.6 (17.7)
Social infrastructure	22.9	27.6	29.6	34.3	41.0	47.4
Total[b]	9,218	23,360	61,958	137,716	340,736	925,471

a. Figures in parentheses: support for small and medium-sized industries.
b. For each period, in 100 million yen.
Source: M. Ogura and T. Yoshino, "Tax System and Treasury Investments and Loans Programme," in R. Komiya, ed., *Industrial Policy in Japan* (University of Tokyo Press, Tokyo, 1984).

transport (shipbuilding), coal-mining and steel, usually through the Japan Development Bank, in the 1953–1955 period (see table 3).

For example, for the planned shipbuilding, financial support with low interest and long-term loans (of more than 15 years) was considered indispensable, because, given the uncertainty of the maritime transport business in that period and its low level of short-term profit, the industry was not able to get the necessary finance from the normal capital market.

The steel industry, for its part, absorbed 15 per cent of Japan Development Bank funds in 1953–1955 to implement the first part of its first plan. For its second plan (1956–1960), it was able to obtain most of the necessary funds from the Long-term Credit Bank and the Industrial Bank of Japan.

In the decade 1965–1975, an important part of the Japan Development Bank's finance was used to strengthen the competitiveness of new industries: large-scale investments were made in the petrochemical industry, automobile industry, automobile part industry, machine-tool industry, electronic industry, etc. These industries absorbed 20 per cent of the total finance of the bank in the period.

Shipbuilding and heavy machinery were also given help to export their products by the Export-Import Bank of Japan. The amount of finance of this bank was approximately half that of the Japan Development Bank in the first half of the 1950s. But this was increased after 1955 as the shipbuilding and heavy machinery industries of Japan became internationally competitive.

Both the Japan Development Bank and the Export-Import Bank of Japan obtain the funds from the Treasury Investments and Loans Programme.

With regard to the second area mentioned in the first section of this paper — that is, small- and medium-sized industries — an important portion of the programme was also channelled to them through government financial institutions.

Approximately 20 per cent of the programme's funds have been used for the modernization of labour-intensive sectors, that is, agriculture and small and medium-sized industries, in the post-war period. Of this 20 per cent, between 13 and 17 per cent was channelled to small and medium-sized industries.

The major public financial institutions through which the funds of the programme are mobilized are: the Small Business Finance Corporation, the People's Finance Corporation, the Shoko Chukin Bank (Central Bank for Associations of Industry and Commerce), the Okinawa Development Corporation, and the Environmental Sanitation Business Finance Corporation. For the agriculture sector, the main financial institution to mobilize the funds of the programme is the Agriculture, Forestry, and Fisheries Finance Corporation.

In the following section, we will discuss in more detail the financial system as it affects small and medium-sized industries in Japan.

The Formation of Financial Institutions for Labour-intensive Sectors

The Japanese experience in the development of financial institutions for labour-intensive sectors (small and medium-sized enterprises and small farmers) seems relevant to contemporary developing countries for the following reasons:

1. The Japanese economy was characterized until quite recently as a labour-surplus economy, and it was necessary to promote the labour-intensive sectors, especially small and medium-sized industries, for the three important reasons mentioned in the first section of this paper.
2. The imperfection of the market mechanism as far as these industries were concerned has been resolved largely by the formation and development of appropriate financial systems, either directly through governmental financial institutions or indirectly through private institutions that are supported by certain policy measures, without this affecting the advantages of the free-market mechanism. In other words, there was created a unique combination of the principle of free competition in the capital market and the policy of mobilizing the necessary funds for the labour-intensive sector through complementary mechanisms.

3. With regard to the agricultural sector, agricultural co-operatives and government financial institutions played an important role in supporting small labour-intensive family farms and, in this way, complemented the market mechanism without affecting its efficiency.

In this section the author will refer to certain Latin American examples, but, as the information related to them is very limited, any such references should be viewed as tentative.

To begin with, we will give an overview of the financial system for small and medium-sized enterprises in Japan. The agricultural sector will be dealt with later.

Financial institutions that mobilize funds for these enterprises may be grouped into the following three categories (see figure 1): (a) private commercial banks which are not specialized in financing small and medium-sized enterprises; (b) private financial institutions which are specially organized to finance these enterprises; (c) government financial institutions which supplement the private institutions' financing of these enterprises, especially in the area of long-term credit.

As is shown in table 4, 53 per cent of the outstanding loans and discounts to these enterprises comes from institutions belonging to group a, while 34 per cent comes from group b and 13 per cent from group c. As far as loans for investment in machinery and equipment are concerned, 49 per cent are from group a, 30 per cent from group b and 22 per cent from group c (as of March 1983).

The three groups have the following structure and characteristics: group a is composed of 13 city banks with 3,027 offices, 63 local banks with 6,114 offices, 3 long-term credit banks with 70 offices and 7 trust banks with 364 offices. As these banks develop their financial activities according to the free-market mechanism, it is natural that they give higher priority to large enterprises because of their high solvency; to smaller enterprises they not only give less priority but also demand higher levels of interest to cover their higher risks and reduced economies of scale. Therefore, in allocating finances commercial banks operate a kind of two-tier system that depends on the size of the enterprise:

1. Banks limit the amount of funds for smaller enterprises when there is a shortage of funds, particularly in periods of high economic growth. Smaller enterprises commonly receive a higher percentage of funds from these banks during periods of recession.
2. There was a considerable difference between interest rate applied to large enterprises and that applied to small and medium-sized enterprises. The difference oscillated between 1 and 2 percentage points in a certain period. (This takes into account the real interest rate for smaller businesses – they were often asked to deposit a certain amount at the same time that they were taking out a loan at a higher interest rate than that applied to large enterprises.)

The institutions which belong to group b are as follows: 71 mutual loan and

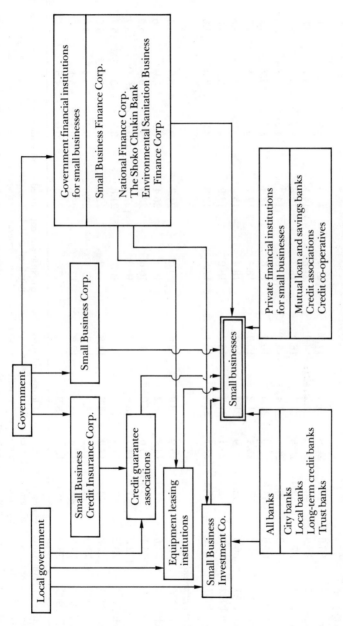

Fig. 1. Japan: Primary financing measures for small business (after Small Business Finance Corporation, *Small Business Financing in Japan and the Small Business Finance Corporation*, SBFC, Tokyo, 1983).

Table 4. Japan: Share in outstanding loans and discounts for business funds of small businesses by type of financial institution (percentages)

	Private financial institutions	All banks	City banks	Local banks	Mutual loan and savings banks	Credit associations	Government financial institutions	Small Business Finance Corp.	People's Finance Corp.	Shoko Chukin Bank	Total
Share of total											
March 1976	86.6	47.4	20.9	21.5	17.9	21.3	13.4	3.9	3.3	5.6	100
March 1977	85.9	46.1	20.3	21.5	18.0	21.8	14.1	4.1	3.4	5.8	100
March 1978	87.2	52.9	23.5	21.3	16.0	18.3	12.8	3.7	3.2	5.3	100
	(85.3)	(46.3)	(20.5)	(21.4)	(18.2)	(20.8)	(14.7)	(4.2)	(3.7)	(6.1)	(100)
March 1979	87.6	53.8	24.0	21.5	16.0	17.8	12.4	3.5	3.2	5.1	100
March 1980	87.3	53.0	23.4	21.2	15.9	18.4	12.7	3.7	3.4	4.9	100
March 1981	86.8	52.4	22.9	21.2	16.0	18.4	13.2	4.0	3.7	4.8	100
March 1982	86.8	52.6	23.1	21.3	16.2	18.0	13.2	4.1	3.7	4.8	100
March 1983	87.2	53.4	23.7	21.4	16.1	17.6	12.8	3.9	3.5	4.8	100
Share of equipment funds											
March 1976	75.7	44.1	15.9	18.0	12.6	19.0	24.3	9.7	5.6	6.4	100
March 1977	73.6	41.9	14.7	18.2	12.3	19.4	26.4	10.9	6.0	6.6	100
March 1978	76.1	48.4	16.6	18.0	11.1	16.6	23.9	9.5	5.6	5.9	100
	(72.8)	(41.2)	(14.7)	(18.6)	(12.6)	(19.0)	(27.2)	(10.9)	(6.4)	(6.7)	(100)
March 1979	78.2	49.5	17.7	19.0	12.0	16.7	21.8	8.2	5.3	5.4	100
March 1980	77.8	48.8	17.2	18.8	11.8	17.2	22.2	8.6	5.2	5.6	100
March 1981	77.0	48.2	17.2	18.7	11.7	17.1	23.0	9.2	5.3	5.6	100
March 1982	77.0	48.0	17.4	18.8	12.0	17.0	23.0	9.4	5.1	5.6	100
March 1983	78.2	48.6	18.3	18.9	12.3	17.2	21.8	9.0	4.6	5.6	100

a. Figures of "All banks" represent the amounts of loans and discounts to corporations capitalized at 50 million yen or less and unincorporated enterprises (as for wholesale and retail trade and services, to corporations capitalized at 10 million yen or less and unincorporated enterprises) until September 1973, to corporations capitalized at 100 million yen or less and unincorporated enterprises (as for wholesale trade, to corporations capitalized at 30 million yen or less and unincorporated enterprises, as for retail trade and services, to corporations capitalized at 10 million yen or less and unincorporated enterprises) from October 1973 to March 1977 and thereafter to corporations capitalized at 100 million yen or less or employing 300 regular workers or less (as for wholesale trade, to corporations capitalized at 30 million yen or less or employing 100 regular workers or less, as for retail trade and services, to corporations capitalized at 10 million yen or less or employing 50 regular workers or less) and unincorporated enterprises.
b. Figures in parentheses are loans and discounts for small business on the preceding definition of "All banks" loans to small business.

Source: Small Business Finance Corporation, *Financial Statistics Quarterly* (Tokyo).

savings banks with 4,096 offices, 456 credit associations with 6,246 offices, and 468 credit co-operatives with 2,682 offices. Although both mutual loan and savings banks (*sogo* banks) and credit associations (*shinyo kinko*) have functions similar to local banks, they finance principally small and medium-sized enterprises and unincorporated enterprises. The mutual loan and savings banks mainly finance enterprises with capital of not more than 400 million yen or with 300 employees or less. The credit associations are non-profit co-operative organizations, whose members consist of residents, workers, and small businessmen in a specific area. Their lending is limited to member small firms with capital of not more than 400 million yen or with 300 employees or less. There is also a ceiling on loans from these institutions: 20 per cent of equity capital or 1,500 million yen, whichever is less, in the case of mutual loan and savings banks: and 20 per cent of equity capital or 800 million yen, whichever is less, in the case of credit associations.

Finally, credit co-operatives are non-profit co-operatives for financing small and petty enterprises. Their lending is limited to members of the co-operatives and to 20 per cent of their equity capital or 400 million yen, whichever is less.

The formation and development of these financial institutions specialized in supporting small and medium-sized enterprises is the result of a balanced combination of market mechanisms and of policies aimed at promoting enterprises that are handicapped in the capital market. From this point of view, the following factors stand out:

First, it was the free-market mechanism that produced the fundamental conditions of development for private financial institutions specializing in small and medium-sized enterprises. As was mentioned earlier, commercial banks set a higher interest rate for smaller enterprises than for large ones. The situation permitted the formation of a different type of smaller institution whose cost of financial intermediation was higher than that of commercial banks, because of, among other things, scale demerit of lending.

Second, the government's basic policy of promoting and protecting these financial institutions for smaller enterprises complemented the above-mentioned conditions of their coexistence with commercial banks. Their financial activities were protected against competition from the commercial banks by the governmental policy of restricting the latter's activities, and especially by limiting their number of branches. However, these financial institutions were required to attend principally to smaller enterprises, the total amount of loans extended to customers other than small and medium-sized firms being restricted to 20 per cent of the total loans outstanding. As far as credit associations were concerned, the total amount of loans extended to non-members was restricted to 20 per cent of their total loans outstanding.

Third, the tradition of mutual assistance among small enterprises in Japan also played an important role. Most of the specialized financial institutions for these enterprises were established and developed by the enterprises themselves.

Their origin is in mutual financing associations, called *mujin* in Japanese, which have a fairly long history. The members of credit associations or credit co-operatives, or even the customers of mutual loan and savings banks who deposit their savings, feel confident that they will get money when they need it from these institutions, while they would not feel this about commercial banks. The difference between the financial institutions and commercial banks could be only relative, but it is important for small enterprises.

The institutions which belong to group c are as follows: the Small Business Finance Corporation (established in 1953) with 58 offices, the People's Finance Corporation (1936) with 149 offices, the Shoko Chukin Bank (1936) with 93 offices, and the Environmental Sanitation Business Finance Corporation (1967).

All these governmental institutions were established to supplement the facilities available to small businesses, especially to supply them with the long-term credits that private institutions find it difficult to extend.

The Small Business Finance Corporation played a particularly important role in the implementation of the government policy of promoting small and medium-sized enterprises. It provides funds for investment in plant and equipment and long-term operating funds, and is particularly useful to enterprises in specific industrial fields that are in line with the government's industrial policy objectives. On the other hand, the National Financial Corporation provides funds exclusively to small businesses.

The Shoko Chukin Bank is a semi-governmental organization whose capital is subscribed by the government and the bank's affiliated organizations, etc. It issues commercial and industrial debenture bonds which are important sources of financial resources.

Finally, the Environmental Sanitation Business Finance Corporation supplies those businesses having to do with environmental sanitation with the funds necessary for raising the level of public sanitation and for modernizing their equipment.

In addition to these special financial institutions, as classified in groups b and c, there are other types of institutions that have been set up to overcome the financial disadvantages of smaller enterprises. The most important is the Small Business Credit Insurance Organization, a governmental institution established in 1958. This entity insures the guarantees of the local credit guarantee associations established in each prefecture and in major cities, thus insuring the repayment of debts by small businesses. These associations were established because of the difficulty which such businesses experience, often due to insufficient collateral.

On the other hand, three Small Business Investment Companies were established in 1963, with contributions from the Small Business Finance Corporation, prefectural governments, and private organizations, in order to supply capital funds to small businesses that find it difficult to raise funds from the capital market or have a limited capacity for capital increase. These companies help with the financing of small businesses, until they are able to raise funds from the capital market by themselves, by providing an investment and consultation service. The Equipment Leasing Institutions were established in 46 prefectures with the contributions from the Small Business Finance Corporation and prefectural governments. They, through equipment leasing systems, allow small businesses that have difficulty in procuring funds for equipment investment to modernize their equipment.

Regarding the financial system for family farms, the major institutions are as follows:

There are 4,373 multi-purpose agricultural co-operatives and 4,294 single-purpose agricultural co-operatives in Japan. The former carry out financial activities, in addition to other activities related to agriculture. They are affiliated to the Prefectural Credit Federations and Central Bank for Agricultural Co-operatives.

The financial institutions for the agricultural and fishery sectors are the Credit Federation of Agricultural Co-operatives, the Credit Federation of Fishery Co-operatives, and the Mutual Insurance Federation of Agricultural Co-operatives.

The total amount of deposits and savings of agricultural co-operatives, and of deposits, savings, and debentures of the Central Bank for Agricultural Co-operatives, is approximately 7.5 per cent of the total amount of principal funds and investments of the financial institutions in Japan (see figure 2).

In addition to these private financial institutions established specially for these sectors, the government provides funds to them through the Agriculture, Forestry, and Fisheries Finance Corporation.

Now, let us compare the system and scale of financial institutions for small and medium-sized enterprises in Japan with those of some major countries of Latin America. We should bear in mind that the figures are not strictly comparable, and therefore the comparison is of a preliminary nature.

In Japan, the percentage of loans outstanding to small and medium-sized enterprises was 47.4 per cent of the total amount of loans outstanding to all enterprises at the end of 1984 (47.2 per cent at the end of 1983). As far as the manufacturing industries are concerned, the percentage of small and medium-sized industries in the total loans outstanding was 66.8 per cent (64.8 per cent at the end of 1983). This last figure is very high if we compare it with the share of such enterprises in the total value added of the manufacturing industries (56.9 per cent in 1980). In Japan manufacturing industries absorbed 19.1 per cent of total loans outstanding to all industries at the end of 1984 (19.9 per cent in 1983).

This contrasts, strongly with the situation in some major Latin American countries. For example, according to an estimate,[4] the percentage of loans outstanding to small and medium-sized enterprises in the manufacturing sector,

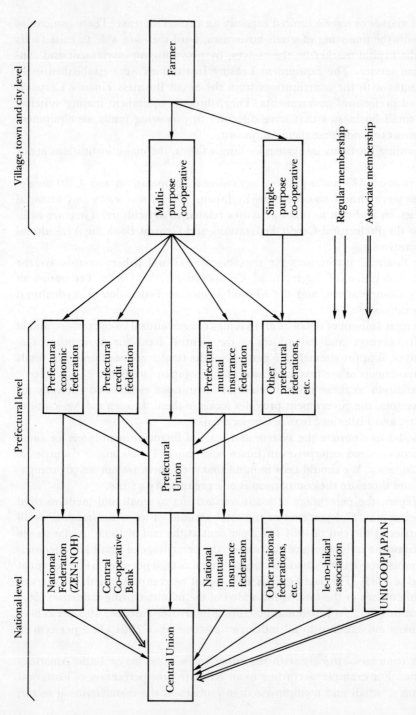

Fig. 2. Japan: Structural outline of agricultural co-operative movement (after Central Union of Agricultural Co-operatives, *The Agricultural Co-operative Movement in Japan*, CUAC, Tokyo, 1984).

compared to total loans outstanding in that sector, was 22.3 per cent in 1979 in Mexico, as against their share in the value-added of 37.8 per cent in 1975. We should add that manufacturing industries absorb approximately 17.0 per cent of total loans outstanding to all industries in Mexico.

In Mexico, it seems that scarcity of finance from formal financial institutions is compensated by enterprises' internal funds or informal financial units (see table 5). In order to give financial support to small and medium-sized enterprises, various governmental programmes were set up, such as Fondo de Garantía y Fomento a la Industria Mediana y Pequeña (FOGAIN) and Programa de Apoyo Integral a la Industria Mediana y Pequeña (PAI).

In Brazil, on the other hand, the Banco Nacional de Desenvolvimiento Economico e Social (BNDES) provides approximately 12 per cent of the funds from all financial institutions. About 20 per cent of the funds of BNDES are used for small and medium-sized enterprises.

Regarding Colombia, according to the Plan Nacional de Desarrollo published in 1983, the following three *credito de fomento* have been used as major instruments for the support of small and medium-sized industries: the Fondo Financiero Industrial (FFI), the Corporación Financiera Popular (CFP), and the Caja de Credito Agrario Industrial y Minero (CA). In addition to them, there are institutions to promote manufacturing industries in general: the Fondo de Inversiones Privadas (FIP), the Fondo de Promoción de Exportaciones (Proexpo), and the Instituto de Fomento Industrial (IFI). Of all loans given to manufacturing industries, more than 80 per cent correspond to credit lines of Proexpo. According to the above-mentioned document, this means the concentration of loans to large enterprises, because 90 per cent of the smaller enterprises (with assets of less than 35 million pesos) do not export. On the other hand, small industries (with assets of less than 60 million pesos) received 25.0 per cent of the *credito de fomento* in 1981.

None of these three countries has liberalized its capital market in recent years, while other major countries of the region, such as Argentine, Chile, Uruguay, Peru, etc., have introduced a liberalization policy. It would be interesting to analyse the change in the participation of small and medium-sized industries in the liberalized capital market in these countries, although unfortunately the author could not obtain the necessary information for this kind of analysis.

Final Remarks: Effects of Liberalization of the Capital Market on Financial Flows for Small and Medium-sized Enterprises in Japan

Although it is very difficult to foresee the effects at this stage, the financial institutions specially organized for small and medium-sized enterprises, as well as the enterprises themselves, will be affected by the liberalization of the capital market in Japan. As for the specialized financial institutions, they will have to

Table 5. Mexico: Source of investment funds for facilities of small- and medium-scale enterprises (percentages)

Kinds of industries	Medium-scale enterprises[a]				Small-scale enterprises[a]			
	Internal funds	Banks	Suppliers of facilities	Others	Internal funds	Banks	Suppliers of facilities	Others
Basic metal	53	35	12	0	43	51	3	3
Metal products	49	30	21	0	46	33	17	4
Non-electrical machinery	55	38	7	0	47	27	20	6
Electrical machinery	38	40	22	0	42	47	11	0
Transport machinery	36	53	10	1	30	62	8	0
Total[b]	46	39	15	0	42	44	12	2
Total for manufacturing industry	52	32	15	1	48	37	11	4

a. Classification according to scale is based upon the value of net worth, which is the criterion taken as a priority reference by FOGAIN for financing purposes. Companies with a net worth of below 5 million pesos (increased to 7 million pesos from the end of July 1980) are considered small-scale enterprises, while those of 5 to 40 million pesos (7 to 60 million pesos from the end of July 1980) are considered medium-scale enterprises. In the present study 72 per cent are small-scale and 28 per cent medium-scale enterprises.
b. Simple averages.
Source: International Development Center (IDC), *A Study on the Development of Manufacturing Industries in the United Mexican States* (IDCJ, Tokyo, 1981).

face stronger competition from commercial banks; and some institutions, especially the smaller mutual loan and savings banks, will have difficulty in competing because of the liberalization of the interest rate among banks and financial institutions. This could weaken the financial system for small and medium-sized enterprises.

On the other hand, we can speculate that enterprises with high solvency, especially medium-sized ones, will be better off because of stronger competition between banks, while other enterprises will have to face less favourable conditions. It is difficult to foresee the impact of this particular phenomenon on the activities of institutions such as the credit associations and credit co-operatives established by small and medium enterprises themselves.

Notes

1. Regarding general aspects of finance, planning and development in Japan, see M. Rietti and A. Hosono, *Planificación y financiamiento del desarrollo: La estrategia de Honduras y la experiencia japonesa* (COFINSA, Tegucigalpa, 1985).
2. K. Ohkawa and M. Tajima, "Small–Medium Scale Manufacturing Industry: A Comparative Study of Japan and Developing Countries," International Development Centre (IDCJ) Working Paper Series, no. A-02 (Tokyo, 1976); S. Motai and K. Ohkawa, "Small-scale Industries: A Study on Japan's 1966 Manufacturing Census," IDCJ Working Paper Series, no. 11 (Tokyo, 1978); A. Hosono, "Industrial Strategy and New Forms of Cooperation," in K. Ohkawa and N. Gonzalez, eds., *Towards New Forms of Cooperation between Latin America and Japan* (ECLA/IDCJ, Santiago/Tokyo, 1980).
3. See T. Yokokura, "Small and Medium Enterprises," in R. Komiya, ed., *Industrial Policy of Japan* (University of Tokyo Press, Tokyo, 1984) (in Japanese).
4. IDCJ, *A Study on the Development of Manufacturing Industries in the United Mexican States* (IDCJ, Tokyo, 1981).

Bibliography

Banco Nacional de Desenvolvimiento Economico e Social (BNDES). *Annual Report 1983*. BNDES, Rio de Janeiro, 1984.

Bank of Japan. *Economic Statistics Annual 1983*. Bank of Japan, Tokyo, 1984.

The Central Union of Agricultural Co-operatives (CUAC). *Agricultural Co-operative Movement in Japan*. CUAC, Tokyo, 1984.

DNP (Unidad de Estudios Industriales). *El sector industrial en el plan nacional de desarrollo*. DNP, Bogotá, 1983.

Jimenez, R. *Organización popular para la producción: Elementos preliminares para la evaluación*. CEPAL, Santiago, 1980.

Small Business Finance Corporation (SBFC). *Small Business Financing in Japan and Small Business Finance Corporation*. SBFC, Tokyo, 1983.

UNIDO. *Actividades de banca para el desarrollo en el decenio de 1980*. UNIDO, Vienna, 1984.

CONTRIBUTORS

Dr. Andrés Bianchi, Chief, Economic Development Division, Economic Commission for Latin America and the Caribbean, Santiago, Chile.

Dr. Akio Hosono, Associate Professor, Institute of Socio-Economic Learning, Tsukuba University, Ibaragi-ken, Japan.

Dr. David Ibarra, Economic Commission for Latin America and the Caribbean, Mexico Office, Mexico, D.F., Mexico.

Dr. Joong-Woong Kim, President, National Information and Credit Evaluation, Inc., Seoul, Republic of Korea.

Dr. Hirohisa Kohama, Associate Professor, University of Shizuoka, Shizuoka-ken, Japan.

Dr. Mario B. Lamberte, Vice-President, Philippine Institute for Development Studies, Metro Manila, Philippines.

Dr. Kazushi Ohkawa, Research and Training Director, International Development Center of Japan, Tokyo, Japan.

Dr. Joseph R. Ramos, Economic Commission for Latin America and the Caribbean, Santiago, Chile.

Dr. Eduardo Sarmiento, Dean of Economics, Universidad de los Andes, Bogotá, Colombia.

Dr. Andrew Sheng, Adviser, Bank Regulation Department, Bank Negara Malaysia, Kuala Lumpur, Malaysia.

CONTRIBUTORS

Dr. Juro Teranishi, Professor, Institute of Economic Research, Hitotsubashi University, Tokyo, Japan.

Dr. Miguel Urrutia, Manager, Economic and Social Development Department, Inter-American Development Bank, Washington, D.C., USA.